T0140960

Guaranteed diagnosis of uncertain linear systems using state-set observation

Dissertation zur Erlangung
des Grades eines Doktor-Ingenieurs
der Fakultät für Elektrotechnik und Informationstechnik
an der Ruhr-Universität Bochum

von Philippe Planchon
aus London, England

Bochum, 2006

1st Reviewer: Prof. Dr.-Ing. Jan Lunze
Ruhr-Universität Bochum, Germany

2nd Reviewer: Prof. Dr.-Ing. Bernd Tibken
Bergische Universität Wuppertal, Germany

Thesis submitted on: 2006-11-22
Date of examination: 2007-02-27

Bibliografische Information der Deutschen Nationalbibliothek

Die Deutsche Nationalbibliothek verzeichnet diese Publikation in der Deutschen Nationalbibliografie; detaillierte bibliografische Daten sind im Internet über http://dnb.d-nb.de abrufbar.

©Copyright Logos Verlag Berlin 2007
Alle Rechte vorbehalten.

ISBN 978-3-8325-1590-4

Logos Verlag Berlin
Comeniushof, Gubener Str. 47,
10243 Berlin
Tel.: +49 030 42 85 10 90
Fax: +49 030 42 85 10 92
INTERNET: http://www.logos-verlag.de

Acknowledgement

This thesis results from five years of research conducted at the institutes of Prof. Jan Lunze both at the Technische Universität Hamburg-Harburg and, for the longest part, at the Ruhr-Universität Bochum.

The present achievement was made possible thanks to the excellent supervision of Prof. Lunze. His profound understanding of dynamic systems has nourished my research activities of interesting theories, most often illustrated by strikingly simple (counter) examples. His strict work ethic and personal commitment are always a source of motivation to me.

This work was greatly supported by ABB Corporate Research Germany. As much financially as by means of proposing an industrial application for which new diagnostic methods had to be studied. Without their support, I would not have had the chance to pursue this research in such good conditions.

I must especially thank two men without whom I would have never started this adventure: Manfred Rode from ABB Corporate Research who was the first to believe in my ability to conduct this work, and Francesco Bullo with whom I discovered my interest in research while at the University of Illinois at Urbana-Champaign.

I wish to thank Prof. Bernd Tibken for accepting to review my thesis and for showing his interest in my research. I also thank Prof. Schmid, Prof. Hofmann and Prof. Tüchelmann for accepting to take part in my examination, as well as Axel, Florian, Jan, Pop and Thorsten for proofreading my manuscript.

On a personal level, many have supported me along the way. First and foremost, my parents, brothers and sisters have all guided and influenced me in their own way. I am grateful to Marion who accepted my expatriation and to Andrea who accompanied me during the final stretch of this work.

My colleagues in Bochum have caused my time in the Ruhrgebiet to be memorable. I think especially of Tobias, Carsten, Christian, Jörg, Pop, Jan, Axel, Florian and many more. The musical and hockey/running friendships have been a great source of distraction. I am also thankful to my childhood friends (Olivier, Philippe, Frédérique, Xavier) as well as to my university friends (Frédéric & Nicola, Sebastian, Jean-François, Christian, Julien, Sylvain, Béa) who have all kept in touch for so long.

Munich, May 2007 Philippe Planchon

La grandeur d'un métier est peut-être, avant tout, d'unir des hommes : il n'est qu'un luxe véritable, et c'est celui des relations humaines.

Antoine de Saint-Exupéry, *Terre des hommes*

Contents

List of Definitions

List of Theorems

List of Propositions

List of Algorithms

Abstract

This thesis presents a model-based diagnostic method used to detect and identify faults in dynamical processes. An observation-based method is studied which considers uncertainties both in the models of the process and in the input and output measurements obtained from the process. Under these conditions the diagnosis is aimed to be guaranteed, *i.e.* it results in a set of faults which always contains the true fault occurring in the process.

The measurement uncertainties are assumed to be unknown-but-bounded, hence at each time instant the input and output vectors are only known to belong to some sets in their respective spaces. Consequently, in pursuing the diagnostic task a set-valued state observer is used to check the consistency of these measurements with the considered models. As each model represents the behaviour of the process for a specific fault, each model consistent with the set-valued measurements accounts for a fault which may have occurred in the process.

In order to achieve the robustness of the diagnosis with respect to model uncertainties, existing algorithms for the state-set observation of linear state-space models are extended in this thesis to the case of time-varying uncertain parameters.

The results obtained, both in the field of guaranteed set-observation and guaranteed diagnosis, are illustrated at the example of a rotating rod. Furthermore, the successful application of this method to an industrial cold rolling mill case-study is shown.

Keywords: diagnosis, unknown-but-bounded uncertainty, uncertain linear systems, time-varying uncertainty, guaranteed estimation, state estimation

Deutsche Kurzfassung

Motivation und Zielsetzung

Die steigende Komplexität moderner Prozesse und Anlagen hat zur Folge, dass aufgetretene Fehler oder Fehlerursachen für einen Anlagebediener schwer erkennbar sind. Wenn die hohe Verfügbarkeit dieser Prozesse einen kritischen Aspekt darstellt, werden neue, automatisierte Diagnoseverfahren benötigt.

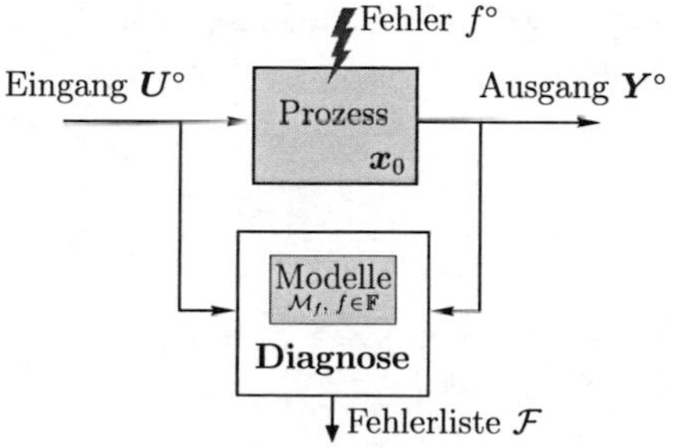

Abbildung 1.: Allgemeine Struktur einer modellbasierten Diagnose

Das Ziel der Diagnose ist der Rückschluss auf die Fehlerursache mittels Ein- und Ausgangsmessungen (Abb. 1). Da viele Fehlerfälle ausschließlich anhand von Veränderungen in den dynamischen Vorgängen erkannt werden können, muss die Diagnose dieses dynamische Verhalten mit berücksichtigen. Ist ein Modell des fehlerfreien Prozesses vorhanden, so kann das gemessene mit dem vom Modell vorhergesagten Verhalten verglichen werden. Stimmen diese nicht überein, so ist ein Fehler erkannt.

Bei ihrer praktischen Umsetzung stoßen *modellbasierte Diagnosemethoden* auf Probleme, die zum Teil durch *Ungenauigkeiten* – sowohl in den gemessenen Signalen als auch in den betrachteten Modellen – verursacht werden. Methoden zur Behandlung von verrauschten Messsignalen sind zwar zahlreich veröffentlicht, benötigen aber meist Angaben über die stochastischen Eigenschaften der Signale. Diese können häufig nur grob approximiert werden oder treffen für einige Störfälle nicht zu. Ein Sensoroffset führt beispielweise zu einem systematischen Messfehler, der nicht stochastisch verteilt ist. Was Modellunsicherheiten betrifft, so ist deren Behandlung in der Praxis begrenzt. Die Darstellung von Unsicherheiten beruht oft auf zusätzlichen Modellelementen, die schwer zu identifizieren sind.

Die vorliegende Arbeit weist folgende Zielstellungen auf:

- Die Betrachtung **systematischer Messfehler**, die keine (bekannten) stochastischen Eigenschaften besitzen.
- Die Verwendung einer **gängigen Modellform**, die Parameterunsicherheiten explizit beinhaltet.
- Die Bestimmung eines **zuverlässigen Diagnoseergebnises**, unter Berücksichtigung der angenommenen Mess- und Modellungenauigkeiten.

Lösungsweg

Es existieren bereits verschiedene Lösungsansätze für die modellbasierte Diagnose. Diese Arbeit beschreibt ein *neues beobachtergestütztes Verfahren.*

Im regelungstechnischen Sinn schätzt ein Beobachter den inneren Prozesszustand anhand gemessener Signale und einem dynamischen Prozessmodell. Für die Diagnose wird ausgenutzt, dass der Schätzfehler des Beobachters gegen Null konvergiert, wenn eine Übereinstimmung zwischen Modell und Prozess besteht.

Die systematischen Messfehler bezüglich der Ein- und Ausgangssignale sind zwar unbekannt, werden jedoch als beschränkt vorausgesetzt. Demzufolge wird der Wert eines Messsignals zu jedem Zeitpunkt durch ein Intervall beschrieben. Diese Betrachtung der Ein- und Ausgangssignale führt zu der Verwendung eines *mengentheoretischen Beobachters.* Dieser schätzt eine Zustandsmenge, die den tatsächlichen inneren Prozesszustand beinhaltet. Aus diesem Grund wird ein derartiger Beobachter auch als gesicherter Beobachter ("guaranteed observer") bezeichnet. Er beruht auf einem Modell in der Zustandsraumdarstellung.

Für den Diagnosezweck wird das Beobachtungsergebnis entsprechend *konsistenzbasierter Prinzipien* ausgewertet. In dieser Arbeit ist bewiesen, dass das für die Beobachtung verwendete Modell nur mit dem Prozess übereinstimmen kann, falls die jeweilige beobachtete Zustandsmenge nicht leer ist. Das Modell wird für diesen Fall als konsistent mit den gemessenen Ein- und Ausgangsignalen bezeichnet.

Dieses Prinzip lässt sich für die Fehleridentifikation auf mehrere Modelle erweitern. Hierzu wird eine Liste $\mathbb{F} = \{f_0, f_1, \ldots, f_N\}$ mit allen möglichen Fehlerfällen vorgegeben. Es wird angenommen, dass der tatsächlich eingetretene Fehler f° zu der Liste gehört und sich zusätzlich jedes Fehlerverhalten im Zustandsraum modellieren lässt. Folgend besteht die Aufgabe der Diagnose darin, die Fehlerfälle zu finden, deren Modelle konsistent mit den Messungen sind.

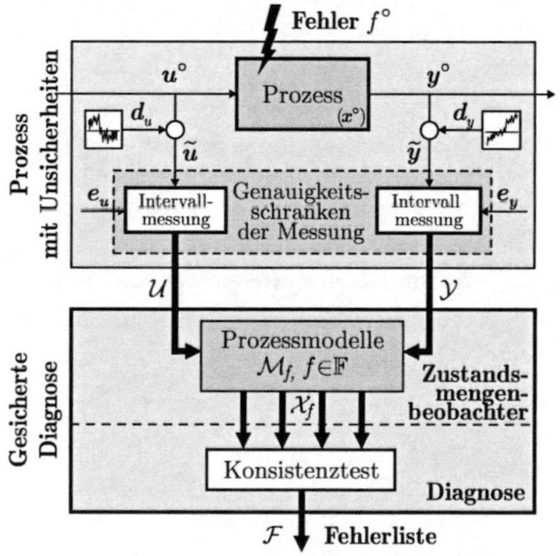

Abbildung 2.: Gesicherte Diagnose mittels Zustandsmengenbeobachtung

Ergebnisse und Beiträge

Die Struktur des in dieser Arbeit konzipierten Diagnoseverfahrens ist in Abb. 2 dargestellt. Das Ergebnis der Diagnose ist eine zeitabhängige Fehlerliste $\mathcal{F}(k) \subset \mathbb{F}$, die stets den tatsächlichen Fehlerfall beinhaltet, d.h. $f^\circ \in \mathcal{F}(k)$.

Das vorgeschlagene Diagnoseverfahren besitzt folgende Vorteile:

- Der Zustandsmengenbeobachter ermöglicht sowohl die Berücksichtigung von systematischen Messfehlern als auch von Prozessmodellen mit Parameterunsicherheit.

- Die Umsetzung des Beobachtungsergebnisses in ein Diagnoseergebnis erfolgt mit Hilfe konsistenzbasierter Prinzipien. Dieses Vorgehen sichert, dass der tatsächliche Fehlerfall immer zum Diagnoseergebnis gehört (*gesicherte Diagnose*).

- Das Verfahren lässt sich in der Praxis einsetzten. Es beruht auf herkömmlichen Zustandsraummodellen und benötigt keine Kenntnisse über die stochastische Verteilung der Messstörsignale.

- Die Robustheit der Methode folgt aus der expliziten Behandlung von Mess- und Modellungenauigkeiten innerhalb des Mengenbeobachters. Daher erfordert diese Diagnosemethode im Vergleich zu anderen Verfahren keine Justierung der Empfindlichkeit gegenüber Unsicherheiten.

Die benötigten Funktionen und Algorithmen für dieses Diagnoseverfahren sind in der SO4CD-Toolbox für MATLAB™ zusammengefasst (Set-Observation for Complete Diagnosis).

Die Diagnosemethode wurde anhand mehrerer numerischer Beispiele getestet. Die praktische Relevanz dieses Diagnoseansatzes ist an einem detaillierten industriellen Beispiel gezeigt.

1. Introduction

A survey of this thesis is given in this chapter. After shortly describing system diagnosis in general and presenting an overview of existing literature dealing with this problem, the content of the thesis is summarised. A the end of this chapter, the notations used in the thesis are presented.

1.1. Diagnostic problem

Modern processes rely increasingly on *automated systems* in order to simplify repetitive tasks while achieving higher processing quality, tighter requirements and an *increased dependability* by avoiding unnecessary human interaction. However, as these systems grow in complexity they are affected by faults with non obvious symptoms: finding the true cause of the failure becomes inextricable for the process operator. A supervision component capable of diagnosing these complex faults is then necessary.

System diagnosis aims at determining whether a process is subject to a fault. This is achieved using the measured inputs and outputs of the process (Fig. 1.1). If specific symptoms are found in the measurements, the diagnostic component determines the presence of a fault and its cause.

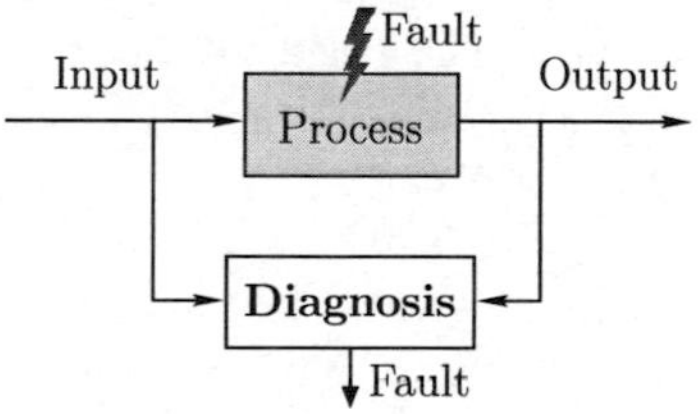

Figure 1.1.: Generic structure of diagnosis

Diagnosis has become an area of active research, because the productivity and availability of complex industrial processes is at stakes. An appropriate supervision avoids costly unplanned down-times. Furthermore, at the example of the automotive industry, the development of new diagnostic methods is pushed forward by laws which require internal processes to operate within a range of acceptable behaviour (*e.g.* to guarantee low emissions). This is only achieved using innovative diagnostic methods to discover faulty components.

The diagnosis is split into three intermediary tasks, [80, 89]:

- **Fault detection** determines *whether a fault occurs.* The exact fault remains unknown.
- **Fault identification** determines *which fault* affects the process.
- **Fault estimation** determines the *magnitude* of the fault affecting the process.

1.2. Literature overview

This overview concentrates on "online" diagnostic methods, *i.e.* methods which are used on a running process. Two classes of methods emerge from the literature: *signal-based* and *model-based* methods. The former encompasses the oldest techniques and is only briefly described here. In-depth analysis of the diagnostic methods discussed below is found in [19, 38, 41, 55, 56, 89, 114–116].

Signal-based diagnosis

Signal-based diagnosis describes supervision methods which aim at finding fault symptoms using pattern recognition techniques. These methods usually see the process as controlled in its operating point and hence only consider its outputs, just as an autonomous system. Methods belonging to this category include

- *Threshold checking.* This describes either *limit checking* in which a signal (*e.g.* temperature or liquid level) is directly monitored to lay within prescribed upper and lower bounds or *change detection* in which a feature is extracted from a signal (*e.g.* variance or derivate) and similarly monitored.

- *Frequency analysis.* This implies a Fast Fourier Transformation (FFT) on a limited time-window of data. The spectral information is used to find fault symptoms.
- *Statistical Process Control (SPC).* Statistical methods are used to compare signals and to extract specific features of the measurements, [48, 51]. *Principal Component Analysis* (PCA, [59]) is a multivariate SPC method which reduces the amount of data to treat in the diagnosis. Statistical diagnosis requires a large data mining to find appropriate attributes to monitor.

Signal-based diagnosis suffers from difficulties encountered to consider dynamic behaviours and to calibrate them (since the symptoms depend on the operating conditions). Furthermore methods treating multiple signals at once are rare and suffer from scalability: their design is difficult to generalise.

Model-based diagnosis

Model-based diagnosis describes supervision methods which aim at finding discrepancies between the actual behaviour of a process (described by its measurements) and its expected behaviour (described by a model). These methods use both the input and the output measurements, such that the dynamic behaviour and the operating conditions are intrinsically considered.

The most distinctive characteristics between these methods is the type of model used. It owes to be *appropriate with respect to the task to be solved*, hence as precise as necessary and as rough as possible [102]. For example, solving a fault estimation task needs precise quantitative information but the detection of the same fault can succeed with a qualitative model. In general the three following model categories are distinguished.

Quantitative models are used in the "Fault Detection and Isolation" (FDI) community. These analytical models are described either in the time-domain as differential equations, for example using a state-space model

$$\dot{\boldsymbol{x}}(t) = \boldsymbol{A}\,\boldsymbol{x}(t) + \boldsymbol{B}\,\boldsymbol{u}(t)$$
$$\boldsymbol{y}(t) = \boldsymbol{C}\,\boldsymbol{x}(t)\ ,$$

or in the frequency-domain as the transfer function relating the Laplace transforms of the input and output signals

$$\boldsymbol{Y}(s) = \boldsymbol{G}(s)\ \boldsymbol{U}(s)\ .$$

The model is said to contain an *analytical redundancy* which allows to diagnose the system. Most quantitative diagnostic methods construct a *residual* $\boldsymbol{r}$ which is (near) zero in the faultless case and larger otherwise. In a perfect situation where models and measurements are exactly known these methods are shown to be equivalent, hence the difficulty lies in obtaining a residual sensitive to the fault occurrence and robust to modelling and measurement errors. Quantitative methods are classified in either of the following.

- **Parity-space approaches** compare the output ($\boldsymbol{y}$) measured at the process with the output ($\tilde{\boldsymbol{y}}$) predicted by the model for the same input. The residual is then defined as $\boldsymbol{r} = \boldsymbol{y} - \tilde{\boldsymbol{y}}$ and a fault is discovered when the residual grows far from the origin $\|\boldsymbol{r}\| \gg 0$, [31,41].

  The model is transformed in order to improve its diagnosis and/or robustness properties. This method suits the detection of additive faults better than multiplicative faults.

- **Observer-based approaches** reconstruct the (non-measurable) inner state of the process using standard methods of control theory. The residual is $\boldsymbol{r} = \boldsymbol{y} - \hat{\boldsymbol{y}}$ where $\hat{\boldsymbol{y}}$ is the output from the observed system, *cf.* [37,39]. For diagnosis, the main effort lies in the appropriate design of the state-observer. The Kalman filter is a well-known estimator which considers measurements subject to stochastic disturbances [23].

  Implementation using bank of observers (*e.g.* Dedicated Observer Scheme, DOS) have shown to be efficient in detecting actuator or sensor faults as well as multiple faults.

- **Parameter identification approaches** aim at finding the parameter value $\hat{\boldsymbol{p}}$ which best suits the measured data, given a specified model structure. Assuming the parameter $\boldsymbol{p}^\circ$ corresponding to the nominal (faultless) model is known, deviations in the residual $\boldsymbol{r} = \boldsymbol{p}^\circ - \hat{\boldsymbol{p}}$ indicate of a faulty behaviour, *cf.* [54].

  This method requires the input and output signals measured during normal process operation to cover a wide enough spectrum such that the dynamic components of the model are correctly identified.

Diagnostic methods based on residual generation can also be used for fault identification by considering the orientation of the vector $\boldsymbol{r}$ in the residual space [41].

Qualitative models emanate from the Artificial Intelligence theory and diagnostic methods using these are referred to as “Diagnosis eXpert systems” (DX community). A qualitative model consists in rough information about a system behaviour and its signals. This implies symbolic descriptions instead of purely numerical values. The advantage of these models is that they correspond to the human way of thinking (*e.g.* a pressure is described as "high", not as equal to 1492.5 [bars]). Because of its coarse nature, a qualitative model is intrinsically robust against measurement and modelling uncertainties, but useful only for severe faults. These methods are classified as follows.

- **Causal-model (or rule-based) approaches** aim at classifying the symptoms and features gained using signal-based methods, [75]. The classification results in rules in the form of "IF *symptom A* AND *symptom B* THEN *fault 1*." Numerous variations of this approach exist such as the consideration of fault probabilities using Bayes networks or the interpretation of the classification rules using Fuzzy theory. The creation of fault trees (hierarchical rules) is automated using *expert systems*.

  The drawback of these methods remains in the inconcise and unclear tables resulting for large systems. Furthermore there is no separation of the physical modelling and the diagnostic algorithm such that the classification effort has to be repeated for each newly considered process.

- **First-principles approaches** compare the behaviour predicted by a qualitative model of the system with its measurements, [47, 66]. The approach is either *consistency-based* or *abductive* depending on the method used for the comparison. The best known implementation is the General Diagnostic Engine (GDE, [33]) in which the process is decomposed in smaller components, each of which is interconnected using qualitative signals.

  The method is efficient as the model knowledge and the diagnostic algorithm are separate, easing the implementation for different processes.

Discrete-event models conveniently describe complex systems in a concise manner, making abstraction of quantitative details such as differential equations (as such they are sometimes seen as qualitative models). These models – also known as formal languages – take one of the following forms.

- **Petri nets** are a common model to describe sequential processes using a transition/node formalism. The nodes describe running subprocesses and the firing of transitions the beginning of new processes. Diagnosis is achieved by detecting the firing of fault transitions and by analysing the reachability of the network, [113]. The research in the field of Petri nets diagnosis is relatively limited, but this model form is known to efficiently detect cycles and deadlocks in parallel-running processes. In [107] a method units Petri nets and automata in a hierarchical diagnostic scheme.

- **Automata** are finite state machines used early on for diagnostic purposes. The states of the automaton are interconnected by transitions which correspond to the occurrence of a specific input and output (an event), [77]. Different solutions to the diagnostic problem exist depending on the formalism assumed for the automata, [36, 97,107].

  An application uses *quantised systems* in which a quantitative model is abstracted to an automaton exhibiting an equivalent qualitative behaviour, [102]. This automaton is simpler to diagnose especially in the case of uncertain or nonlinear processes. While the diagnostic algorithm is cost efficient in its computation, it consumes a lot of storage memory for large systems. Research to deal with this issue considers a modular structural decomposition, [70,87]. An example comparing the diagnosis of quantised system with a semi-qualitative method based on interval models is found in [9].

Approaches to handle uncertainties

Issues caused by uncertainties mainly affect quantitative diagnostic methods. Qualitative methods are rough by definition and hence cope with uncertainties intrinsically.

Making a quantitative diagnosis robust to model or measurement uncertainties implies making either the underlying computations (parity equations, state-observation or parameter identification) robust to uncertainties, or the diagnostic decision (or both). The former approach is referred to as *active* since it considers uncertainties explicitly in the entire diagnostic algorithm, the latter is *passive* as it operates in the final decision stage only (for example using an adaptive threshold when monitoring a residual).

The consideration of uncertainties leads to either optimal or set-valued approaches. There exists few publications comparing these methods, with the notable exception of [17,81,100]. *Optimal methods* aim at finding the "best" solution given a mathematical criterion. In control, the H_2 and H_∞ filtering techniques are frequently used, [40,41,118]. The Kalman filter is an almost standard approach of state-observation, however it is usually limited to a stochastic uncertainty assumption [39,82].

Set-theoretic methods, on the contrary, do not aim at finding a single (supposedly "best") solution but describe all possible solutions. Set methods are well suited to deal with non stochastic errors. They suffer from the engineer's perspective who desires a single solution to any problem, whereas set-valued methods seek the exact solution of the problem which can result in no solution or in many solutions. Set-theoretic methods, however, are the only approaches which can *guarantee statements* in the presence of uncertainties [57]. Using these methods for control is surveyed in [18, 85].

Similar work in the literature

In this thesis a model-based diagnostic method is proposed which combines *consistency-based principles* with a *state-observer* designed to consider both unknown-but-bounded measurement errors and model uncertainties.

Because of the type of uncertainties assumed, a set-valued solution to the observation task is sought. Such observers exist in the literature for state-space models and for different choices of the set's shape: mostly using ellispoids [28, 29, 103] or variants of polytopes [13, 30, 120]. This thesis proposes a polyhedric set-observer which handles *model uncertainties.* To the author's knowledge, the extension of the set-observation for uncertain models has only been recently published for ellispoids and zonotopes [13, 91].

An extensive publication list in the area of diagnosis using interval models is found from the authors of [93–95]. Their work distinguishes itself from the present work in the interpretation of the observation towards diagnosis which is not consistency-based and relies on an input-output model (not a state-space model). Recent results from the group have introduced the use of consistency principles but in the area parameter identification diagnosis [53, 95]. Another research group has used the consistency principles and uncertain models in the frequency-domain [45, 46]. A residual based diagnosis using subpaving sets is studied in [11, 12].

1.3. Aim, assumptions and survey of results

Even when considering a finite set of faults $\mathbb{F}$ which can occur, it is often impossible to determine which single fault $f^\circ \in \mathbb{F}$ – the *true fault* – affects the process. On the diagnostic level, a strategic decision needs to be taken. Either a single fault is pointed out by the diagnosis, or a set of possible faults is described. A diagnosis indicating a single fault has to assume additional information to make its final decision (*e.g.* using statistical knowledge about the frequency of the fault occurrence). In this case, however, the indicated fault may not be the true culprit for the process malfunction and the diagnosis generates a *false alarm.* Alternatively, and this is the strategy followed in this thesis, the diagnosis describes a set $\mathcal{F}^*$ containing *all faults* which can currently occur in the process. These are referred to as *fault candidates.*

The multiplicity of fault candidates – and hence the ambiguity in finding the true fault – has a twofold explanation. First, the considered problem may be *non-diagnosable.* In this case, two or more faults cannot be distinguished given the process measurements. For example, consider a water tank with inflow and water level measurements. Two faults are considered which represent leakages in different areas of the tank's base: obviously, the two faults cannot be distinguished using the available measurements and the problem is said to be non-diagnosable.[1] Second reason for the multiplicity of the fault candidates, the *uncertainties* in the measurements may hide the symptoms of a fault: abrupt changes are covered by noise or a too slow sampling rate, small changes in a sensor value can be caused by an erroneous measurement or a fault, *etc.* Consequently the fault behaviours may not be distinguishable, yielding multiple fault candidates.

Aim of the diagnosis. A diagnostic method is sought which describes *all fault candidates*, *i.e.* all faults for which the models may explain the measured behaviour (*cf.* Def. 2.8). For this purpose an approximation of the measurement uncertainties is used together with multiple models of the process, each characterising a specific fault behaviour. Since the true fault f° affecting the process is a fault candidate, this approach ensures not to oversee the true malfunction of the process. This is referred to as a "*guaranteed diagnosis*" (Def. 2.3).

AIM OF THE DIAGNOSIS	
Given:	– An Input-Output measurement of a process
	– An assumption on the measurement uncertainties
	– The models of the process (faultless and faulty behaviours)
Find:	– The set $\mathcal{F}^*$ of all fault candidates

Assumptions. Throughout the thesis the following assumptions are used:

I. All faults $f \in \mathbb{F}$ which can affect the process are known ("closed-world assumption" [66]) and their respective models $\mathcal{M}_f$ are complete and linear (possibly with parameter uncertainties). The notion of completeness is found in Def. 2.4.

II. A single fault $f^\circ \in \mathbb{F}$ affects the process during the entire time horizon of the diagnosis, hence $f^\circ(k) \equiv f^\circ$.

[1] Referring to the intermediate tasks of diagnosis, the faults are *detectable* but *non-identifiable.* However, if the leakages have different outflow rates, then the faults are identifiable using dynamic models.

III. At all times an upper bound of the measurement error is known. That is, two bounds $\boldsymbol{e}_u(k)$ and $\boldsymbol{e}_y(k)$ are known for which

$$|\boldsymbol{u}^\circ(k) - \tilde{\boldsymbol{u}}(k)| \leq \boldsymbol{e}_u(k) \quad \text{and} \quad |\boldsymbol{y}^\circ(k) - \tilde{\boldsymbol{y}}(k)| \leq \boldsymbol{e}_y(k)$$

holds, with $\boldsymbol{u}^\circ$ the true input and $\tilde{\boldsymbol{u}}$ the measured input (and respectively with $\boldsymbol{y}^\circ$ and $\tilde{\boldsymbol{y}}$ the true and measured outputs).

Main results. A *set-valued representation* of signals is used to diagnose dynamical processes. In this framework, a set-valued state observer is described based on which a consistency-based diagnostic scheme computes the set $\mathcal{F}^*$ of fault candidates. This set contains by definition the true fault f°.

In general, the cost of computing the set of fault candidates is extremely high. Therefore, *guaranteed computation methods* are used which allow to efficiently describe an outer approximation:

$$\boxed{\mathcal{F} \supset \mathcal{F}^*} \, . \tag{1.1}$$

This set contains the true fault at all times, yielding a guaranteed diagnosis (Theorem 9):

$$\boxed{f^\circ \in \mathcal{F}} \, . \tag{1.2}$$

In order to obtain this diagnostic result, the following steps are pursued:

1. *A guaranteed state observer* is used to obtain a set-valued state reconstruction of processes subject to uncertainties (Chapter 3, Theorem 4).

2. *A consistency test* is found which relates the result of the guaranteed observation to the definition of model consistency (Section 4.1, Theorem 7).

3. *A guaranteed diagnostic method* is derived which determines the set of faults $\mathcal{F}(k)$ based on the consistency test. The algorithm guarantees $f^\circ \in \mathcal{F}(k)$ for the considered model and measurement uncertainties (Section 4.2, Theorem 9).

1.4. Structure of thesis

Chapter 2 introduces the basic concepts of diagnosis using a behavioural description of the process. The class of models used to represent the faultless and the faulty behaviours as well as the type of uncertainties considered on the measurements are presented. The consistency principles used for diagnosing such a system are explained leading to the definitions of fault candidates and model consistency. Finally, a description of the diagnostic task pursued for these specifications is given.

Chapter 3 presents the set-valued observation method which will be used to test the consistency of models. The initial-state and final-state observation problems are described for both real-valued and set-valued measurements. This leads to the realisation of a state-set observer which is guaranteed – for the given assumptions – to include the true state of the process. The implementation of the algorithm using polyhedric state-sets is detailed. Results about the observability of a system in this set-theoretic framework conclude the chapter.

Chapter 4 presents the diagnostic method proposed in this thesis. It begins by defining the relationship existing between the state-set observation and the model consistency property. As such it unites the results of the two preceding chapters by describing a guaranteed diagnosis which solves the pursued diagnostic task. Considerations with respect to the diagnosability of dynamical systems conclude the chapter.

Chapter 5 and **Chapter 6** contain numerical examples to which the presented method is applied. The former introduces a motor model which illustrates both the state-set observation and the guaranteed diagnosis. The latter summarises the results from an industrial cooperation which studied the diagnosability of a rolling mill and to which the described diagnosis is applied.

Finally, **Chapter 7** summarises and concludes this thesis. It includes an outlook on possible research topics related to the present contribution.

1.5. Notations

The notations used in this thesis to describe dynamic systems are described in the following. Additionally, a detailed nomenclature is found in Appendix C.

1.5.1. Vectors, Sets and Sequences

Lower-case bold letters refer to *vectors*. The input, output and state vectors are described by $\boldsymbol{u}(k)$, $\boldsymbol{y}(k)$ or $\boldsymbol{x}(k)$ which respectively belong to $\mathbb{R}^m$, $\mathbb{R}^r$ or $\mathbb{R}^n$.

Upper-case calligrahic letters refer to *sets*. The letter corresponds to the contained signal, Sets of input, output and state vectors are noted $\mathcal{U}(k)$, $\mathcal{Y}(k)$ and $\mathcal{X}(k)$.

Furthermore, sequences of values (over time) are considered. A *sequence of vectors* is noted with corresponding upper-case bold letter:

$$\boldsymbol{U}(0\dots k) := (\boldsymbol{u}(0),\dots,\boldsymbol{u}(k))$$
$$\boldsymbol{Y}(0\dots k) := (\boldsymbol{y}(0),\dots,\boldsymbol{y}(k))$$
$$\boldsymbol{X}(0\dots k) := (\boldsymbol{x}(0),\dots,\boldsymbol{x}(k))\ .$$

Sequences of sets are defined similarly. As sets are already written upper-case, the notation distinguishes itself by the number of arguments in the parenthesis:

$$\mathcal{U}(0\dots k) := (\mathcal{U}(0),\dots,\mathcal{U}(k))$$
$$\mathcal{X}(0\dots k) := (\mathcal{X}(0),\dots,\mathcal{X}(k))$$
$$\mathcal{Y}(0\dots k) := (\mathcal{Y}(0),\dots,\mathcal{Y}(k))\ .$$

The concept of an *input-output pair* (I/O pair) is used repeatedly and refers to either of:

- a real-valued I/O pair $(\boldsymbol{u},\boldsymbol{y})(k) = (\boldsymbol{u}(k),\boldsymbol{y}(k))$,
- a sequence of real-valued I/O pairs $(\boldsymbol{U},\boldsymbol{Y})(0\dots k) = \big((\boldsymbol{u},\boldsymbol{y})(0),\dots,(\boldsymbol{u},\boldsymbol{y})(k)\big)$,
- a set-valued I/O pair $(\mathcal{U},\mathcal{Y})(k) = (\mathcal{U}(k),\mathcal{Y}(k))$,
- a sequence of set-valued I/O pairs $(\mathcal{U},\mathcal{Y})(0\dots k) = \big((\mathcal{U},\mathcal{Y})(0),\dots,(\mathcal{U},\mathcal{Y})(k)\big)$.

The time-indexes are left-out when clear from the context. As such, $\boldsymbol{u}$, $\boldsymbol{U}$ and $\mathcal{U}$ are shorthand notations for $\boldsymbol{u}(k)$, $\boldsymbol{U}(0\dots k)$ and either $\mathcal{X}(k)$ or $\mathcal{X}(0\dots k)$. The distinction between the lower and upper cases is therefore primordial: $\boldsymbol{x} \in \mathcal{X}$ refers to one time instant – *i.e.* a vector $\boldsymbol{x}(k)$ belonging to a set $\mathcal{X}(k)$, while $\boldsymbol{Y} \in \mathcal{Y}$ refers to a time horizon – *i.e.* a sequence of vectors $\boldsymbol{Y}(0\dots k)$ belonging to a sequence of sets $\mathcal{Y}(0\dots k)$.

1.5.2. Comparative operators

An inequality comparing two vectors $\boldsymbol{a}$ and $\boldsymbol{b}$ is read component-wise:

$$\boldsymbol{u} \le \tilde{\boldsymbol{u}} \qquad \Longleftrightarrow \qquad u_i \le \tilde{u}_i,\ \forall i \in \{1,\dots,m\}\ .$$

Two sequences of same length are equal if all their elements are equal:

$$\boldsymbol{U}_1(0\ldots\bar{k}) = \boldsymbol{U}_2(0\ldots\bar{k}) \quad\Longleftrightarrow\quad \boldsymbol{u}_1(k) = \boldsymbol{u}_2(k),\ \forall k \in \{0,\ldots,\bar{k}\}\ .$$

A sequence of vectors is included in a sequence of sets if the inclusion holds for all elements:

$$\boldsymbol{U}(0\ldots\bar{k}) \in \mathcal{U}(0\ldots\bar{k}) \quad\Longleftrightarrow\quad \boldsymbol{u}(k) \in \mathcal{U}(k),\ \forall k \in \{0,\ldots,\bar{k}\}\ .$$

1.5.3. System operators ψ and Ψ

State-space models $\mathcal{M}_f$ are considered to describe the behaviour of processes. Hence, a process subject to an input ($\boldsymbol{u}$) generates an internal state ($\boldsymbol{x}$) and a measured output ($\boldsymbol{y}$). For a given model, the path of reaction from a (real-valued) input sequence to its outputs is formalised using the *input-to-output operators* $\psi_{y,f}$ (for an output value) and $\Psi_{y,f}$ (for an output sequence). Similarly, the *input-to-state operators* $\psi_{x,f}$ and $\Psi_{x,f}$ are defined.

These operators represent the simulation of model $\mathcal{M}_f$ from the perspective of an initial state $\boldsymbol{x}(0)$. Hence, the input-to-output operator is a time-domain equivalent of the transfer function $G(s) = \frac{Y(s)}{U(s)}$ used in the frequency-domain, but considers the initial state. Similar operators exist in the literature, *e.g.* the "measured output function" in [17] or the "transformation" $\mathcal{T}$ in [36].

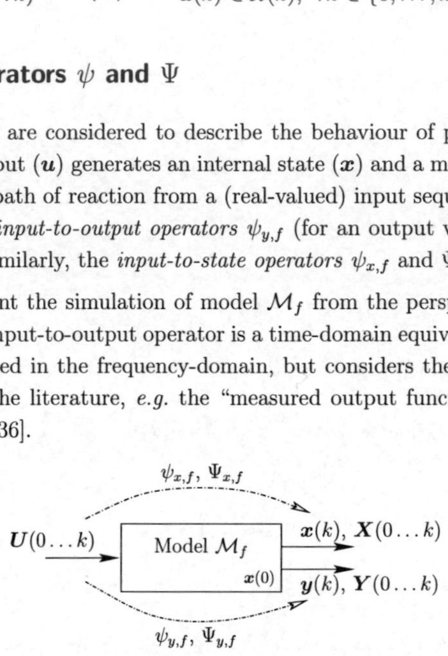

Both exact models (noted $\mathcal{M}_f(\boldsymbol{p})$) and uncertain models (noted $\mathcal{M}_f(\mathbb{P})$) are considered. The result of their corresponding operators is either real-valued or set-valued. If necessary, this distinction is made by the additional subscript "$\boldsymbol{p}$" or "$\mathbb{P}$."

- Case of exact models $\mathcal{M}_f \equiv \mathcal{M}_f(\boldsymbol{p})$:

$$\begin{aligned}
\Psi_{y,f}\big(\boldsymbol{x}(0),\ \boldsymbol{U}(0\ldots k)\big) &= \boldsymbol{Y}(0\ldots k) &&:= \Psi_{y,f,\boldsymbol{p}}\big(\boldsymbol{x}(0),\ \boldsymbol{U}(0\ldots k)\big)\\
\psi_{y,f}\big(\boldsymbol{x}(0),\ \boldsymbol{U}(0\ldots k)\big) &= \boldsymbol{y}(k) &&:= \psi_{y,f,\boldsymbol{p}}\big(\boldsymbol{x}(0),\ \boldsymbol{U}(0\ldots k)\big)\\
\text{and } \Psi_{x,f}\big(\boldsymbol{x}(0),\ \boldsymbol{U}(0\ldots k-1)\big) &= \boldsymbol{X}(0\ldots k) &&:= \Psi_{x,f,\boldsymbol{p}}\big(\boldsymbol{x}(0),\ \boldsymbol{U}(0\ldots k-1)\big)\\
\psi_{x,f}\big(\boldsymbol{x}(0),\ \boldsymbol{U}(0\ldots k-1)\big) &= \boldsymbol{x}(k) &&:= \psi_{x,f,\boldsymbol{p}}\big(\boldsymbol{x}(0),\ \boldsymbol{U}(0\ldots k-1)\big)\ .
\end{aligned}$$

- Case of unknown models $\mathcal{M}_f \equiv \mathcal{M}_f(\mathbb{P})$:

$$\begin{aligned} \Psi_{y,f}\big(\boldsymbol{x}(0),\ \boldsymbol{U}(0\ldots k)\big) &= \mathcal{Y}(0\ldots k) &&:= \Psi_{y,f,\mathbb{P}}\big(\boldsymbol{x}(0),\ \boldsymbol{U}(0\ldots k)\big) \\ \psi_{y,f}\big(\boldsymbol{x}(0),\ \boldsymbol{U}(0\ldots k)\big) &= \mathcal{Y}(k) &&:= \psi_{y,f,\mathbb{P}}\big(\boldsymbol{x}(0),\ \boldsymbol{U}(0\ldots k)\big) \\ \text{and } \Psi_{x,f}\big(\boldsymbol{x}(0),\ \boldsymbol{U}(0\ldots k-1)\big) &= \mathcal{X}(0\ldots k) &&:= \Psi_{x,f,\mathbb{P}}\big(\boldsymbol{x}(0),\ \boldsymbol{U}(0\ldots k-1)\big) \\ \psi_{x,f}\big(\boldsymbol{x}(0),\ \boldsymbol{U}(0\ldots k-1)\big) &= \mathcal{X}(k) &&:= \psi_{x,f,\mathbb{P}}\big(\boldsymbol{x}(0),\ \boldsymbol{U}(0\ldots k-1)\big)\ . \end{aligned}$$

The operators are written for the general case in which a direct feedthrough might make the output $\boldsymbol{y}(k)$ dependent on $\boldsymbol{u}(k)$. Therefore, the input-to-output operators require a longer input sequence than the input-to-state operators. The explicit computations of these operators is given in Section 2.2.

This thesis considers uncertain models with time-varying parameters. To describe the (exact) model corresponding to a specific sequence of parameters $\boldsymbol{P} = \boldsymbol{P}(0\ldots k) = (\boldsymbol{p}(0),\ldots,\boldsymbol{p}(k)) \in \mathbb{P}^{k+1}$, the "$\boldsymbol{P}$" subscript is appended instead of "$\mathbb{P}$." Intuitively it is clear that the decomposition

$$\psi_{y,f,\mathbb{P}}\big(\boldsymbol{x}(0),\boldsymbol{U}(0\ldots k)\big) := \bigcup_{\boldsymbol{P}\in\mathbb{P}^{k+1}} \psi_{y,f,\boldsymbol{P}}\big(\boldsymbol{x}(0),\boldsymbol{U}(0\ldots k)\big) = \mathcal{Y}(0\ldots k)$$

holds, and similarly for the other operators of the uncertain model.

2. Diagnosis of dynamic systems

This chapter describes the concepts underlying consistency-based diagnosis. After introducing the behaviour of systems, the classes of model and measurement uncertainties to be considered are described. For these classes, the chosen consistency test is detailed and finally the two main steps of the pursued diagnostic approach are given, leading way to the following Chapters 3 and 4.

2.1. Behavioural description

Diagnosing a process aims at determining what fault is affecting the process using its input and output measurements. Additionally, model-based methods use a description of the expected process behaviour for the different faulty situations $f \in \mathbb{F}$ (Fig. 2.1). An approach to tackle this problem based on the behavioural concepts of [119] is explained in the following.

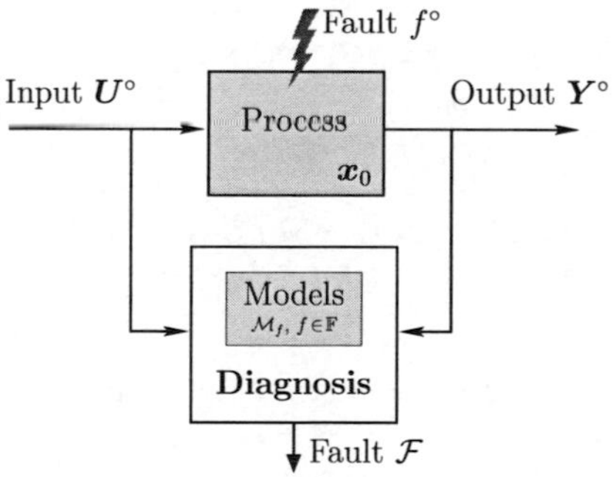

Figure 2.1.: Generic structure of model-based diagnosis

The system behaviour is defined as follows.

Definition 2.1 (Universum and System Behaviour) *The* universum $\mathbb{U}$ *describes the space in which the input and output vary. If sequences of any length* $k \in \mathbb{N}$ *are considered for the inputs* $\boldsymbol{u} \in \mathbb{R}^m$ *and outputs* $\boldsymbol{y} \in \mathbb{R}^r$*, the universum is then* $\mathbb{U} = (\mathbb{R}^m \times \mathbb{R}^r)^{\mathbb{N}}$.
The subset of the universum which may occur for a given system is its behaviour $\mathcal{B}^\circ$. *The behaviour describes those sequences of input and output which are compatible with the laws governing the considered system.*

The concept underlying this definition is illustrated in Fig. 2.2. The space spanned by the $\boldsymbol{U}$ and $\boldsymbol{Y}$ axis represents the universum, a subset of which is depicted as the behaviour $\mathcal{B}^\circ$. For this system, one measured I/O sequence pair $(\boldsymbol{U}_A, \boldsymbol{Y}_A)(0 \ldots \bar{k})$ is represented by the point A.

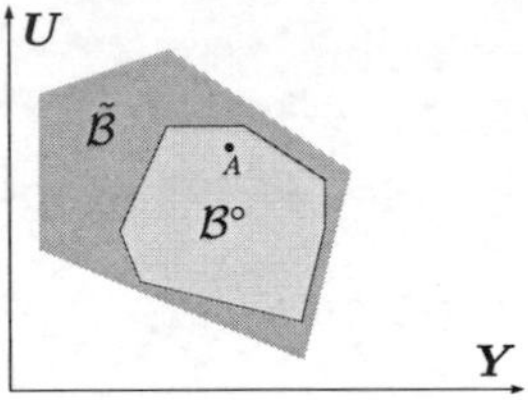

Figure 2.2.: Universum and system behaviour

Assumption II says that the system is subject to a single fault during the considered time horizon. Therefore, it is possible to separately consider the different faults, leading to the description of multiple behaviours $\mathcal{B}_f^\circ$ for each fault $f \in \mathbb{F}$. Based on this formalism, the ideal diagnosis is to describe the set $\mathcal{F}^\circ$ defined as follows.

Definition 2.2 (True Set of Faults $\mathcal{F}^\circ$) *The true set of faults is the set of all faults, which system behaviours contain the I/O sequence* $(\boldsymbol{U}^\circ, \boldsymbol{Y}^\circ)(0 \ldots k)$*:*

$$\mathcal{F}^\circ(k) := \{f \in \mathbb{F} \mid (\boldsymbol{U}^\circ, \boldsymbol{Y}^\circ)(0 \ldots k) \in \mathcal{B}_f^\circ\} \, .$$

Proposition 2.1 (True fault in the true set of faults) *The true set of faults contains the true fault:* $f^\circ \in \mathcal{F}^\circ(k)$, $\forall k \in \mathbb{N}$. ■

The above result follows directly from the definition and needs not be proven. Given the fault behaviours $\mathcal{B}_f^\circ$, $f \in \mathbb{F}$, the true set of faults depends on the scenario described by the I/O, as well as on the length $(k+1)$ of the given sequence.

The fact that the true set of faults may not be limited to a single element, in which case $\mathcal{F}^\circ \supsetneq \{f^\circ\}$, emphasises the problems related to *diagnosability* of the system: some I/O sequences do not permit to distinguish different system behaviours.

Definition 2.3 (Complete and Guaranteed Diagnosis)
A diagnostic algorithm resulting in a set of faults $\mathcal{F}$ is said to be complete *iff $\mathcal{F}^\circ \subset \mathcal{F}$, and is* guaranteed *iff $f^\circ \in \mathcal{F}$.*

Hence, a complete diagnosis describes more faults than could actually have occurred. These faults $f \in \mathcal{F} \backslash \mathcal{F}^\circ$ are referred to as *spurious solutions.* Furthermore, from Prop. 2.1 follows that a complete diagnosis is also guaranteed, since $f^\circ \in \mathcal{F}^\circ \subset \mathcal{F}$.

Definition 2.4 (Complete Behaviour, Sound Behaviour) *The behaviour $\mathcal{B}_{\text{complete}}$ is said to be a complete behaviour of $\mathcal{B}^\circ$ iff*

$$\mathcal{B}^\circ \subseteq \mathcal{B}_{\text{complete}} \ .$$

Similarly, the behaviour $\mathcal{B}_{\text{sound}}$ is said to be a sound behaviour of $\mathcal{B}^\circ$ iff $\mathcal{B}_{\text{sound}} \subseteq \mathcal{B}^\circ$.

Modelled behaviour. In practice, the system behaviour $\mathcal{B}_f^\circ$ is not precisely known and cannot be stored on a computer based on Def. 2.1. Therefore, a *mathematical model* $\mathcal{M}_f$ of the system is used to generate a *modelled behaviour* $\tilde{\mathcal{B}}_f \approx \mathcal{B}_f^\circ$. For example, using the input-to-output operator $\Psi_{y,f}$ (*cf.* preliminary "Notations" chapter), the behaviour of a model $\mathcal{M}_f$ is generated from

$$\tilde{\mathcal{B}}_f := \{(\boldsymbol{U}, \boldsymbol{Y})(0 \dots k) \mid \exists \boldsymbol{x}_0 \in \mathbb{R}^n,\ \boldsymbol{Y}(0 \dots k) \in \Psi_{y,f}(\boldsymbol{x}_0, \boldsymbol{U}(0 \dots k)),\ k \in \mathbb{N}\} \ . \tag{2.1}$$

Due to inexact system parameters, approximate structural representation or even voluntary simplification of the model $\mathcal{M}_f$, the modelled behaviour $\tilde{\mathcal{B}}_f$ it generates only approximates the system behaviour $\mathcal{B}_f^\circ$. In this thesis, models are assumed to generate a complete behaviour, *i.e.* $\mathcal{B}_f^\circ \subseteq \tilde{\mathcal{B}}_f$. Such models are referred to as *complete models* of the system.[1]

2.2. Process models

Dynamic process models are used to simulate an approximate behaviour of the process and/or to design and test controllers which have to comply with strict specifications on the system's dynamic behaviour. *Diagnostic models* are used to verify whether the measurements of a process (its I/O signals) comply with the expected behaviour of the process (its model). Discrete-time linear state-space diagnostic models are used in this thesis, which may or may not contain information about model uncertainties.

[1]This notion of completeness is taken from [102], and not from aforementioned [119].

Exactly known models

The well-known discrete-time linear state-space representation

$$\mathcal{M}_f: \quad \boldsymbol{x}(k+1) = \boldsymbol{A}_f\,\boldsymbol{x}(k) + \boldsymbol{B}_f\,\boldsymbol{u}(k) \tag{2.2}$$

$$\boldsymbol{y}(k) = \boldsymbol{C}_f\,\boldsymbol{x}(k) + \boldsymbol{D}_f\,\boldsymbol{u}(k) \tag{2.3}$$

is chosen to represent the process behaviour under influence of a fault $f \in \mathbb{F}$. Here $\boldsymbol{x} \in \mathbb{R}^n$, $\boldsymbol{u} \in \mathbb{R}^m$ or $\boldsymbol{y} \in \mathbb{R}^r$ respectively stand for the state, input or output vectors while $\boldsymbol{A}_f$, $\boldsymbol{B}_f$, $\boldsymbol{C}_f$ or $\boldsymbol{D}_f$ stand for the system, the input, the output or the direct feedthrough matrices (with according dimensions). Although the method proposed in this thesis can be used with time-varying models, the lengthier notations $\boldsymbol{A}_f(k)$, $\boldsymbol{B}_f(k)$, $\boldsymbol{C}_f(k)$ and $\boldsymbol{D}_f(k)$ are left out. Furthermore the system matrix $\boldsymbol{A}_f$ is assumed to be regular.[2] This is a common assumption for set-valued methods, [28]. It is rarely an issue with physical systems for which $\boldsymbol{A}_f$ is obtained from the discretisation of a continuous-time system, [5].

The above model is a mathematical representation of the process obtained for specific values of some physical process parameters

$$\boldsymbol{p} = \left(p_1, \ldots, p_{N_p}\right)^T .$$

To emphasise this relation, the exact model $\mathcal{M}_f$ may be written as $\mathcal{M}_f(\boldsymbol{p})$ with

$$\mathcal{M}_f(\boldsymbol{p}): \quad \boldsymbol{x}(k+1) = \boldsymbol{A}_f(\boldsymbol{p})\,\boldsymbol{x}(k) + \boldsymbol{B}_f\,\boldsymbol{u}(k) \tag{2.4}$$

$$\boldsymbol{y}(k) = \boldsymbol{C}_f\,\boldsymbol{x}(k) + \boldsymbol{D}_f\,\boldsymbol{u}(k) . \tag{2.5}$$

The input-to-state operator $\psi_{x,f}$ for exact state-space models $\mathcal{M}_f(\boldsymbol{p})$ has a well-known analytic form:

$$\begin{aligned} \psi_{x,f}\big(\boldsymbol{x}_0, \boldsymbol{U}(0\ldots\bar{k}-1)\big) &= \psi_{x,f,\boldsymbol{p}}\big(\boldsymbol{x}_0, \boldsymbol{U}(0\ldots\bar{k}-1)\big) \\ &= \boldsymbol{A}_f(\boldsymbol{p})^{\bar{k}}\boldsymbol{x}_0 + \sum_{j=1}^{\bar{k}} \boldsymbol{A}_f(\boldsymbol{p})^{\bar{k}-j}\boldsymbol{B}_f\boldsymbol{u}(j-1) \\ &= \boldsymbol{x}(\bar{k}) . \end{aligned} \tag{2.6}$$

The description of the input-to-output operator $\psi_{y,f}$ is straightforward using the above result and Eq. (2.5) and so are the sequence operators $\Psi_{x,f}$ and $\Psi_{y,f}$ by repeating the above computation for all time instants of the sequence horizon.

[2]The inverse of $\boldsymbol{A}_f$ is needed in Section 3.4, *cf.* Lemma 3.1 or Prop. 3.2.

Models with uncertainties

As the robustness of diagnostic methods against uncertainties is primordial, these should be considered within the diagnostic model. The focus is laid on uncertainties affecting the system matrix $\boldsymbol{A}_f$. This eases the solution of the diagnostic problem while considering the most important source of model uncertainties. Furthermore, uncertainties in the matrices $\boldsymbol{B}_f$ and $\boldsymbol{D}_f$ may be considered alternatively, for example using additional uncertainties on the input signal $\boldsymbol{u}$ (*cf.* Section 2.3). Therefore, from all uncertain linear processes, only those with uncertainties in the output matrix $\boldsymbol{C}_f$ cannot be considered using the proposed diagnostic method.

In the literature, [10,58], two methods to describe uncertainties in state-space models are found:

- *Structured uncertainty*, or "physically motivated uncertainty."
  This type of uncertainty considers the system matrix as a function of the process parameters p_i:

  $$\boldsymbol{A}_f = \boldsymbol{A}_f(\boldsymbol{p}) = \boldsymbol{A}_{f,0} + \sum_{i=1}^{N_p} (p_i \, \boldsymbol{A}_{f,i})$$

  with $\boldsymbol{A}_{f,i}$ constant matrices and

  $$p_i \in [p_i^{\min},\, p_i^{\max}], \quad i \in \{1, \dots, N_p\},$$

  the uncertain parameters of the process.

- *Unstructured uncertainty*, or "mathematically motivated uncertainty."
  This type of uncertainty considers the system matrix as the sum of two components

  $$\boldsymbol{A}_f = \boldsymbol{A}_{f,0} + \boldsymbol{\Delta}_{A_f}$$

  with $\boldsymbol{A}_{f,0}$ constant and $\boldsymbol{\Delta}_{A_f}$ an uncertain matrix constrained by the positive definite matrix $\boldsymbol{Q}_{f,0} \succ 0$ as

  $$\boldsymbol{\Delta}_{A_f}^T \boldsymbol{\Delta}_{A_f} \leq \boldsymbol{Q}_{f,0} \, .$$

In this thesis, a structured model uncertainty is considered because it is the closest from the underlying engineering applications. Indeed, models are abstracted from differential equations using specific process parameters. It is the uncertainties in these parameters which should be described in the model.

In the following, the model with uncertainties is noted $\mathcal{M}_f(\mathbb{P})$ and is represented by the time-varying uncertainties model

$$\mathcal{M}_f(\mathbb{P}): \quad \boldsymbol{x}(k+1) = \boldsymbol{A}_f(\boldsymbol{p}(k))\,\boldsymbol{x}(k) + \boldsymbol{B}_f\,\boldsymbol{u}(k) \tag{2.7}$$

$$\boldsymbol{y}(k) = \boldsymbol{C}_f\,\boldsymbol{x}(k) + \boldsymbol{D}_f\,\boldsymbol{u}(k) \tag{2.8}$$

$$\boldsymbol{p}(k) \in \mathbb{P} \tag{2.9}$$

with the structured decomposition of the system matrix

$$\boldsymbol{A}_f(\boldsymbol{p}(k)) = \boldsymbol{A}_{f,0} + \sum_{i=1}^{N_p} (p_i(k)\,\boldsymbol{A}_{f,i}) \,. \tag{2.10}$$

The uncertainties are described by the intervals within which the components p_i of $\boldsymbol{p}$ vary:[3]

$$\mathbb{P} := \{\boldsymbol{p} \in \mathbb{R}^{N_p} \mid p_i^{\min} \leq p_i \leq p_i^{\max}, \quad i \in \{1, \ldots, N_p\}\}. \tag{2.11}$$

Clearly, the uncertain model $\mathcal{M}_f(\mathbb{P})$ from Eqs. (2.7)–(2.9) is equivalent to an exact model $\mathcal{M}_f(\boldsymbol{p})$ from Eqs. (2.4)–(2.5) when

$$\mathbb{P} = \{\boldsymbol{0}\} \quad \text{and} \quad \boldsymbol{A}_{f,0} := \boldsymbol{A}_f(\boldsymbol{0}) \,. \quad \text{(And } \boldsymbol{A}_{f,i},\ i \geq 1 \text{, may take any value.)}$$

The input-to-state operator $\psi_{x,f}$ for uncertain state-space models $\mathcal{M}_f(\mathbb{P})$ is set-valued and hence cannot be expressed as simply as its exact model counterpart in Eq. (2.6). However the following holds

$$\begin{aligned} \psi_{x,f}\big(\boldsymbol{x}_0, \boldsymbol{U}(0 \ldots \bar{k}-1)\big) &= \psi_{x,f,\mathbb{P}}\big(\boldsymbol{x}_0, \boldsymbol{U}(0 \ldots \bar{k}-1)\big) \\ &= \bigcup_{\boldsymbol{P} \in \mathbb{P}^{\bar{k}}} \Big(\psi_{x,f,\boldsymbol{P}}\big(\boldsymbol{x}_0, \boldsymbol{U}(0 \ldots \bar{k}-1)\big)\Big) \\ &= \mathcal{X}(\bar{k}) \,, \end{aligned} \tag{2.12}$$

where $\psi_{x,f,\boldsymbol{P}}$ is the input-to-state operator of the exact model corresponding to a specific sequence of parameters $\boldsymbol{P} = \boldsymbol{P}(0 \ldots \bar{k}-1) = (\boldsymbol{p}(0), \ldots, \boldsymbol{p}(\bar{k}-1)) \in \mathbb{P}^{\bar{k}}$. An analytical expression is found in [15] as:

$$\begin{aligned} \psi_{x,f,\boldsymbol{P}}\big(\boldsymbol{x}_0, \boldsymbol{U}(0 \ldots \bar{k}-1\big) = &\left(\prod_{k=0}^{\bar{k}-1} \boldsymbol{A}_f\big(\boldsymbol{p}(k)\big)\right) \boldsymbol{x}_0 \\ &+ \sum_{j=1}^{\bar{k}} \left(\prod_{q=j}^{\bar{k}-1} \boldsymbol{A}_f\big(\boldsymbol{p}(q)\big)\right) \boldsymbol{B}_f\ \boldsymbol{u}(j-1) \,, \end{aligned} \tag{2.13}$$

[3] Although $\mathbb{P}$ is a set, the calligraphic notation $\mathcal{P}$ is not used because it is a *fixed* set and does not vary with time, as for example the input sets $\mathcal{U}(k)$ or state sets $\mathcal{X}(k)$ do.

which relies on the following matrix product definitions

$$\text{for } k' > k: \quad \prod_{q=k}^{k'} \boldsymbol{A}_f(\boldsymbol{p}(q)) := \boldsymbol{A}_f(\boldsymbol{p}(k')) \cdot \boldsymbol{A}_f(\boldsymbol{p}(k'-1)) \cdot \ldots \cdot \boldsymbol{A}_f(\boldsymbol{p}(k))$$

$$\text{for } k' = k-1: \quad \prod_{q=k}^{k-1} \boldsymbol{A}_f(\boldsymbol{p}(q)) := \boldsymbol{I} \, .$$

The descriptions of the input-to-output operator $\psi_{y,f}$ and the sequence operators $\Psi_{x,f}$ and $\Psi_{y,f}$ are directly derived from this description of $\psi_{x,f}$.

Remark 2.1 – For a constant parameter (*i.e.* a time-invariant sequence $\boldsymbol{P} = (\boldsymbol{p}, \ldots, \boldsymbol{p})$), the two Eqs. (2.13) and (2.6) are equivalent and $\psi_{x,f,\boldsymbol{p}} \equiv \psi_{x,f,\boldsymbol{P}}$. Therefore, a model with uncertain but time-invariant parameters is seen as a set of models with $\mathcal{M}_f(\mathbb{P}) = \{\mathcal{M}_f(\boldsymbol{p}) \mid \boldsymbol{p} \in \mathbb{P}\}$. If the system behaviour corresponds to a specific realisation of this state-space model for a constant parameter vector $\boldsymbol{p}^\circ \in \mathbb{P}$ then the uncertain model is complete since $\mathcal{M}_f(\boldsymbol{p}^\circ) \in \mathcal{M}_f(\mathbb{P})$ which implies that the modelled behaviour contains the system behaviour.

Remark 2.2 – Although the uncertain model is referred to as "uncertain linear model" or "linear model with parametric uncertainties," the considered model is actually nonlinear. This terminology is nevertheless used in the literature, for example with interval linear systems [57, 117].

2.3. Measurement uncertainties

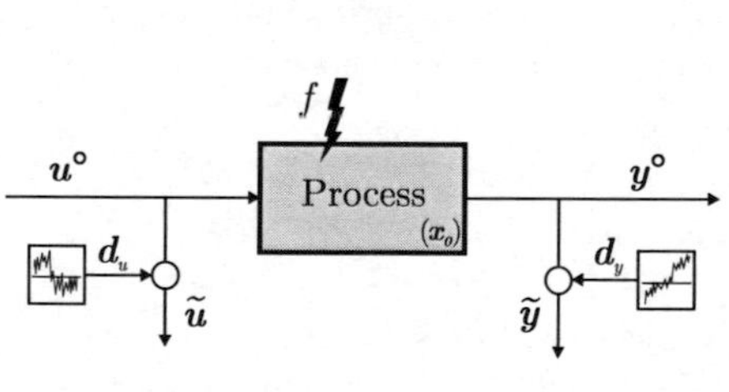

Figure 2.3.: Erroneous input and output measurements

Until now, the input $\boldsymbol{u}^\circ$ and output $\boldsymbol{y}^\circ$ signals affecting the process were assumed to be exactly known. In practice, as shown in Fig. 2.3, the measurements of the true I/O signals are affected by the measurement disturbances $\boldsymbol{d}_u$ and $\boldsymbol{d}_y$. The measured input and output differ from their true values, such that

$$\tilde{\boldsymbol{u}}(k) = \boldsymbol{u}^{\circ}(k) + \boldsymbol{d}_u(k) \neq \boldsymbol{u}^{\circ}(k)$$
$$\tilde{\boldsymbol{y}}(k) = \boldsymbol{y}^{\circ}(k) + \boldsymbol{d}_y(k) \neq \boldsymbol{y}^{\circ}(k) \ .$$

In order to obtain a reliable diagnostic result, the measurement uncertainties $\boldsymbol{d}_u$ and $\boldsymbol{d}_y$ need to be considered. In practice, the measurement disturbance is considered to be either one of the following:

- stochastically distributed,
- unknown-but-bounded,
- energy-bounded.

The *stochastic* consideration of measurement uncertainties are appropriate when the signals $\tilde{\boldsymbol{u}}$ and $\tilde{\boldsymbol{y}}$ are subject exclusively to noise. The disturbances $\boldsymbol{d}_u$ and $\boldsymbol{d}_y$ are described using stochastic properties such as those of white or coloured noise. The stochastic uncertainties are then handled using optimal approaches [100], a well-known example of which is the Kalman filter in the field of state observation [23].

The *unknown-but-bounded* measurement uncertainties consider the disturbance signals $\boldsymbol{d}_u$ and $\boldsymbol{d}_y$ to be unknown except for an upper bound in their amplitude

$$|\boldsymbol{d}_u(k)| \leq \boldsymbol{e}_u(k) \quad \text{and} \quad |\boldsymbol{d}_y(k)| \leq \boldsymbol{e}_y(k) \ . \tag{2.14}$$

Such a disturbance is not supposed to possess specific stochastic properties and may not be energy-bounded [16, Chap. 5]. It is the only assumption which appropriately considers errors caused for example by wrong sensor calibration, measurement offsets, *etc.*

The *energy-bounded* disturbance assumes that

$$\int_0^{+\infty} \boldsymbol{d}_u(t)\,\mathrm{d}t < +\infty \quad \text{and} \quad \int_0^{+\infty} \boldsymbol{d}_y(t)\,\mathrm{d}t < +\infty \ .$$

In [17] the authors consider the state observation under both "energy-type" (energy bounded) and "instantaneous-type" (unknown-but-bounded) of uncertainty constraints and shows an analogy in the obtained solution steps (for the energy constraint approach) with the optimal Kalman observation steps.

In the case of unknown or energy bounds, the frequency domain H_2 and H_∞ filtering techniques are most often used [34, 40, 118, 121].

In this thesis such **unknown-but-bounded measurement uncertainties** are considered using a set-theoretic framework. For this purpose, the real-valued input and output measurements $\tilde{\boldsymbol{u}}(k)$ and $\tilde{\boldsymbol{y}}(k)$ are converted to set-valued measurements $\mathcal{U}(k)$ and $\mathcal{Y}(k)$ using the operators Γ_u and Γ_y:

$$\mathcal{U}(k) = \Gamma_u\big(\tilde{\boldsymbol{u}}(k), \boldsymbol{e}_u(k)\big) = \big\{\boldsymbol{u} \in \mathbb{R}^m \mid |\boldsymbol{u} - \tilde{\boldsymbol{u}}(k)| \leq \boldsymbol{e}_u(k)\big\} \tag{2.15}$$

$$\mathcal{Y}(k) = \Gamma_y\big(\tilde{\boldsymbol{y}}(k), \boldsymbol{e}_y(k)\big) = \big\{\boldsymbol{y} \in \mathbb{R}^r \mid |\boldsymbol{y} - \tilde{\boldsymbol{y}}(k)| \leq \boldsymbol{e}_y(k)\big\} . \tag{2.16}$$

Such measurement sets are also referred to as *interval measurements* (or interval sets), since they are constructed from component-wise inequalities. An example of such a set is shown in Fig. 2.4.

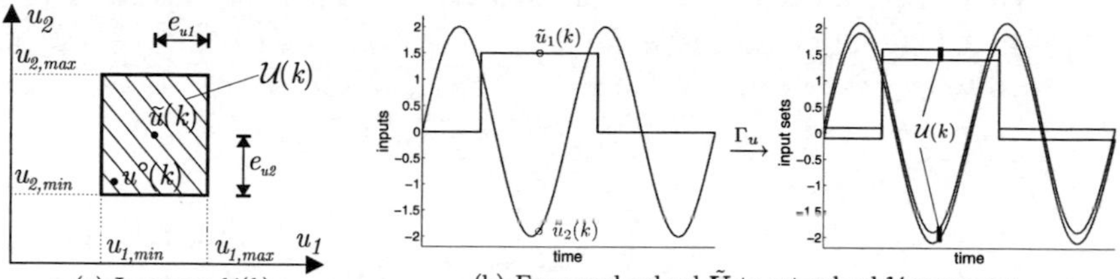

(a) Input set $\mathcal{U}(k)$

(b) From real-valued $\tilde{\boldsymbol{U}}$ to set-valued $\mathcal{U}$ sequences

Figure 2.4.: Measurement set at the example of a two-dimensional input $\boldsymbol{u} = (u_1, u_2)^T$.

Computations involving the set-valued input $\mathcal{U}(k)$ and output $\mathcal{Y}(k)$ amount to more complex tasks then those only involving the real-valued $\tilde{\boldsymbol{u}}(k)$ and $\tilde{\boldsymbol{y}}(k)$. However, assuming the bounds described in the Inequalities (2.14) hold at all time k (Assumption III), then:

$$\boldsymbol{u}^\circ(k) \in \mathcal{U}(k)$$
$$\text{and } \boldsymbol{y}^\circ(k) \in \mathcal{Y}(k) .$$

Based on this, a diagnostic method which considers the whole set-valued measurements $\mathcal{U}(k)$ and $\mathcal{Y}(k)$ leads to a guaranteed diagnostic result, *i.e.* a result which includes the true fault f° (Def. 2.3).

Definition 2.5 (Admissible Sequence) *A sequence*

$$\boldsymbol{U}(0\ldots\bar{k}) = \big(\boldsymbol{u}(0), \ldots, \boldsymbol{u}(\bar{k})\big)$$

is said to be admissible over the sequence of sets $\mathcal{U}(0\ldots\bar{k}) = \big(\mathcal{U}(0), \ldots, \mathcal{U}(\bar{k})\big)$ *iff*

$$\boldsymbol{u}(k) \in \mathcal{U}(k), \ \forall k \in \{0, \ldots, \bar{k}\} .$$

The notation $\boldsymbol{U}(0\dots\bar{k}) \in \mathcal{U}(0\dots\bar{k})$ *is introduced to denote an admissible sequence. The set sequence* $\mathcal{U}$ *is said to* contain *the real-valued sequence* $\boldsymbol{U}$.

Using the notion of admissible sequence, Assumption III is equivalently formulated as "the sequences of true input and output $(\boldsymbol{U}^{\circ}, \boldsymbol{Y}^{\circ})$ are admissible over the sequences of measured input and output sets $(\mathcal{U}, \mathcal{Y})$."

2.4. Consistency-based diagnosis

In the following the principles of consistency checking are introduced, based on which the consistency of real and set-valued signals with models is described. These principles lead to the complete (and hence guaranteed) diagnostic method developed in this thesis.

Definition 2.6 (Input-Output Consistency) *The I/O sequence* $(\boldsymbol{U}, \boldsymbol{Y})(0\dots\bar{k})$ *is said to be consistent with the model* $\mathcal{M}_f$ *if it is contained in the modelled behaviour,* i.e. $(\boldsymbol{U}, \boldsymbol{Y})(0\dots\bar{k}) \in \tilde{\mathcal{B}}_f$. *This relation between model and I/O sequence is noted*

$$\mathcal{M}_f \models (\boldsymbol{U}, \boldsymbol{Y})(0\dots\bar{k}) \,.$$

The input-output consistency can also be described using the input-to-output sequence operator $\Psi_{y,f}$. Considering the behavioural set $\tilde{\mathcal{B}}_f$ as in Eq. (2.1), it follows that for known models

$$\mathcal{M}_f(\boldsymbol{p}) \models (\boldsymbol{U}, \boldsymbol{Y})(0\dots\bar{k}) \iff \exists \boldsymbol{x}_0 \in \mathbb{R}^n, \text{ s.t. } \boldsymbol{Y}(0\dots\bar{k}) = \Psi_{y,f,\boldsymbol{p}}(\boldsymbol{x}_0, \boldsymbol{U}(0\dots\bar{k}))$$

and for uncertain models

$$\mathcal{M}_f(\mathbb{P}) \models (\boldsymbol{U}, \boldsymbol{Y})(0\dots\bar{k}) \iff \exists \boldsymbol{x}_0 \in \mathbb{R}^n, \text{ s.t. } \boldsymbol{Y}(0\dots\bar{k}) \in \Psi_{y,f,\mathbb{P}}(\boldsymbol{x}_0, \boldsymbol{U}(0\dots\bar{k})) \,.$$

Remark 2.3 – It is said that *the state* $\boldsymbol{x}_0$ *is consistent with the sequence of I/O* $(\boldsymbol{U}, \boldsymbol{Y})$ (with respect to the model $\mathcal{M}_f$) because its existence is necessary to the consistency of the I/O with the model.

Definition 2.7 (Input-Output Set Consistency)
The I/O set sequence $(\mathcal{U}, \mathcal{Y})(0\dots\bar{k})$ *is said to be consistent with the model* $\mathcal{M}_f$ *if there exists at least one admissible I/O sequence in the modelled behaviour,* i.e.

$$\exists (\boldsymbol{U}, \boldsymbol{Y})(0\dots\bar{k}) \in (\mathcal{U}, \mathcal{Y})(0\dots\bar{k}), \;\; \textit{s.t. } \mathcal{M}_f \models (\boldsymbol{U}, \boldsymbol{Y})(0\dots\bar{k}) \,.$$

Analogously to Def. 2.6, this relation between model and I/O set sequence is noted

$$\mathcal{M}_f \models (\mathcal{U}, \mathcal{Y})(0 \ldots \bar{k}) \ .$$

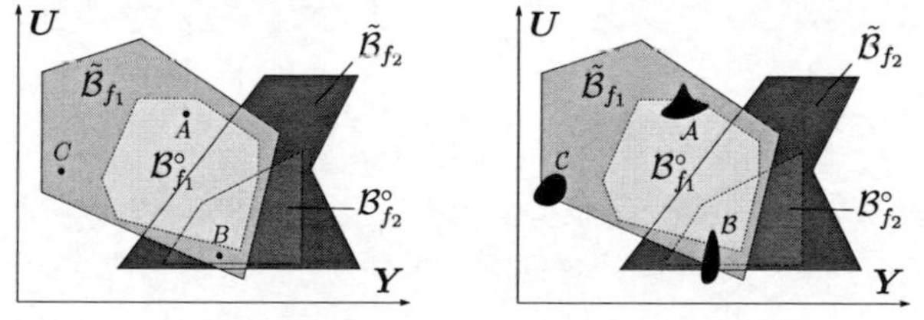

(a) Case of real-valued I/O (black points A, B, C)

(b) Case of set-valued I/O (black sets $\mathcal{A}$, $\mathcal{B}$, $\mathcal{C}$)

Figure 2.5.: Behavioural sets and I/O measurements

An example is given in Fig. 2.5 to illustrate the above consistency definitions. In the figure, both the system behaviours ($\mathcal{B}^{\circ}_{f_1}$ and $\mathcal{B}^{\circ}_{f_2}$ for two faults f_1 and f_2) as well as the modelled behaviours ($\tilde{\mathcal{B}}_{f_1}$ and $\tilde{\mathcal{B}}_{f_2}$ generated for two complete models $\mathcal{M}_{f_1}$ and $\mathcal{M}_{f_2}$ of the faults) are drawn.

In the real-valued case, the I/O sequences are shown as black points in the universum (left figure). From Def. 2.6 all sequences A, B and C are consistent with $\mathcal{M}_{f_1}$, only B is consistent with $\mathcal{M}_{f_2}$. Differently in the set-valued case, the I/O set sequences are shown as black areas in the universum (right figure). The consistency from Def. 2.7 then holds when these areas intersect the modelled behaviours in *at least one point*. Therefrom all drawn set sequences $\mathcal{A}$, $\mathcal{B}$ and $\mathcal{C}$ are consistent with $\mathcal{M}_{f_1}$, only $\mathcal{B}$ is consistent with $\mathcal{M}_{f_2}$.

This illustrates an important aspect of consistency-based diagnosis: despite its name, it is the *inconsistencies* with modelled behaviours which are of significance for the diagnosis [19]. Indeed, the set consistency definition requires that at least one admissible sequence be consistent with the modelled behaviour (see *e.g.* $\mathcal{A}$). Unfortunately, it can never be proven that the (unknown) true I/O sequence belongs to the modelled behaviour and even less to the system behaviour. However, assuming the true I/O sequence is admissible over the measured set $\mathcal{A}$, the inconsistency of $\mathcal{A}$ with model $\mathcal{M}_{f_2}$ proves the inconsistency of the true I/O sequence with $\mathcal{M}_{f_2}$ (and hence with the system behaviour $\mathcal{B}_{f_2}$ if the model is complete). The measurement $\mathcal{A}$ *cannot* correspond to fault f_2.

Definition 2.8 (Set of Fault Candidates $\mathcal{F}^*$)
The set of fault candidates is the set of all faults, for which the models are consistent with the considered I/O set sequence:

$$\mathcal{F}^*(\bar{k}) := \{f \in \mathbb{F} \mid \mathcal{M}_f \models (\mathcal{U},\mathcal{Y})(0\ldots\bar{k})\} .$$

Theorem 1 (Complete models lead to a complete set of fault candidates)
If for all considered faults $f \in \mathbb{F}$, complete models $\mathcal{M}_f$ are known, then the resulting set of fault candidates contains the true set of faults: $\mathcal{F}^\circ(\bar{k}) \subset \mathcal{F}^*(\bar{k})$. The set of fault candidates describes consequently a complete and guaranteed diagnosis.

PROOF: For all $f \in \mathcal{F}^\circ(\bar{k})$, $(\boldsymbol{U}^\circ, \boldsymbol{Y}^\circ)(0\ldots\bar{k}) \in \mathcal{B}_f^\circ \subset \tilde{\mathcal{B}}_f$. As the true I/O is assumed admissible over the measured set sequences $(\mathcal{U},\mathcal{Y})$, then $\mathcal{M}_f \models (\mathcal{U},\mathcal{Y})(0\ldots\bar{k})$, hence $f \in \mathcal{F}^*(\bar{k})$. The final sentence follows directly from Def. 2.3. □

By analogy to Def. 2.2, the set of fault candidates is equivalently described as

$$\mathcal{F}^*(\bar{k}) = \{f \in \mathbb{F} \mid \exists(\boldsymbol{U},\boldsymbol{Y}) \in (\mathcal{U},\mathcal{Y})(0\ldots\bar{k}), \text{ s.t. } (\boldsymbol{U},\boldsymbol{Y})(0\ldots\bar{k}) \in \tilde{\mathcal{B}}_f\}$$

where $\tilde{\mathcal{B}}_f$ is the behaviour generated by the model $\mathcal{M}_f$.

Using the above definitions to interpret Fig. 2.5, then $\mathcal{F}^* = \{f_1\}$ in the case of the I/O set $\mathcal{A}$ and $\mathcal{F}^* = \{f_1, f_2\}$ for the I/O set $\mathcal{B}$. For the purpose of example, suppose the real-valued I/O sequences noted by the points A and B are the true I/O (with $A \in \mathcal{A}$ and $B \in \mathcal{B}$). From Fig. 2.5(a), the true set of faults is $\mathcal{F}^\circ = \{f_1\}$ for the former and $\mathcal{F}^\circ = \{f_2\}$ for the latter case. This illustrates that if $f \in \mathcal{F}^*$ it cannot be assured that $f \in \mathcal{F}^\circ$. E contrario, if a fault f exists, such that $f \notin \mathcal{F}^*$, it is proven that $f \notin \mathcal{F}^\circ$ (and hence $f \neq f^\circ$ as of Prop. 2.1). Therefore consistency-based diagnostic methods rest upon the *exclusion of faults* in determining their result.

While the true set of faults $\mathcal{F}^\circ$ represents an "ideal diagnosis" which requires knowledge of the *system behaviours* ($\mathcal{B}_f^\circ$) for each fault, the set of fault candidates $\mathcal{F}^*$ represents the "best possible diagnosis" given the *models representing the faults* and the considered *measurement uncertainties*. Therefore, process diagnosis can only aim at reconstructing the set of fault candidates $\mathcal{F}^*$. An example of this is seen looking back at Fig. 2.5: the set-valued sequence $\mathcal{C}$ is consistent with the modelled behaviour $\tilde{\mathcal{B}}_{f_1}$ (hence $f_1 \in \mathcal{F}^*$) but not with the system behaviour $\mathcal{B}_{f_1}^\circ$ (hence $f_1 \notin \mathcal{F}^\circ$): the model of f_1 is too coarse to exclude the fault from the set of candidates.

Remark 2.4 – The term "consistency" used in this thesis is quite common in the diagnostic literature [19, 47, 53, 96]. The term of "model validation" found in [98] describes a similar concept in which consistent models are "validated models." Opposed to these principles of consistency-based diagnosis, an "abductive diagnosis" is also described in [5, 75]. This approach uses sound approximations of the system behaviour and therefore cannot achieve a guaranteed diagnostic result.

2.5. Guaranteed diagnosis approach

In the introductory chapter a general "Aim of the diagnosis" is described. Having detailed the type of measurements and models to be used within a consistency-based diagnostic framework, the following task is considered in this thesis:

DIAGNOSTIC TASK	
Given:	– Set-valued Input-Output measurements $\mathcal{U}(0\ldots k)$ and $\mathcal{Y}(0\ldots k)$ which contain the true I/O – Complete models $\mathcal{M}_f$ for each fault $f \in \mathbb{F}$ (modelled behaviour contains the system behaviour)
Find:	– A set $\mathcal{F}(k)$ containing all fault candidates ($\mathcal{F} \supseteq \mathcal{F}^*$)

A solution to the above task is a **guaranteed diagnosis** since the resulting set $\mathcal{F}(k)$ contains the true fault at *all time k*. In order to solve the task, given the models of the fault behaviours and the set-valued I/O measurements, a method to check the consistency as defined in Def. 2.7 is necessary.

This thesis describes a solution to this task. The resulting structure of the diagnosis is shown in Fig. 2.6 and consists of two main steps:

1. A *guaranteed state observation* computes a set of states $\mathcal{X}_f(k)$ which can occur under consideration of the I/O sets and the model $\mathcal{M}_f$ (Chapter 3).

2. A *consistency test* uses the result of the guaranteed observation to verify consistency and derive therefrom the set of faults $\mathcal{F}$ (Chapter 4).

This method can compute the set of fault candidates, in which case $\mathcal{F} \equiv \mathcal{F}^*$. However, this is only possible in special cases and is computationally expensive (*cf.* Theorem. 10).

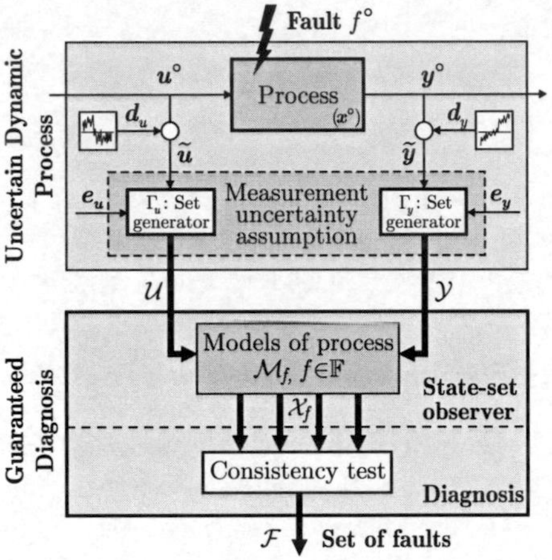

Figure 2.6.: Guaranteed diagnosis using state-set observation

3. Guaranteed state observation

A guaranteed state observer is presented based on set-valued measurements of the input and output. It computes the set of all states consistent with this type of measurement. This chapter concentrates on describing the state-set observation and its properties. The following Chapter 4 applies this observer towards consistency-based diagnosis.

Remark 3.1 – This chapter considers a single model ($\mathcal{M}_f$) at all times. Nevertheless, the subscript "f" is kept in order to preserve the notations throughout the thesis. Furthermore, this underlines which variables depend on the model (*e.g.* $\boldsymbol{x}_f^*$, $\mathcal{X}_f$, ...).

3.1. State observation task

This section describes the (real-valued) state observation task for an exact model $\mathcal{M}_f \equiv \mathcal{M}_f(\boldsymbol{p})$. The proposed framework is then extended in full length for uncertain measurements and uncertain models in the next section leading to a guaranteed (hence set-valued) state observation.

Independently of the considered framework, the aim of state observation is similar in all areas of control engineering. A formulation thereof is proposed as follows.

STATE OBSERVATION TASK (Fig. 3.1)	
Given:	– Input-Output sequences $\boldsymbol{U}(0\ldots\bar{k})$ and $\boldsymbol{Y}(0\ldots\bar{k})$
	– The state-space model $\mathcal{M}_f$ of the process
Find:	– The state of the process $\boldsymbol{x}_f^\circ(0)$ of $\boldsymbol{x}_f^\circ(k)$

The state observation task is separated into two distinct problems depending on the time instant of interest, hence the terms of *initial-state* and *final-state* observation problems.

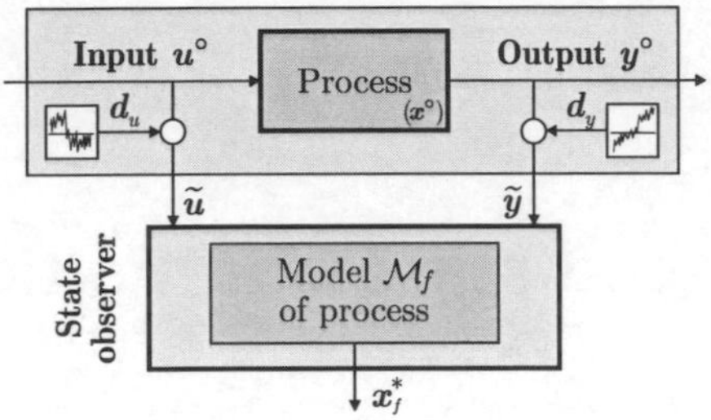

Figure 3.1.: State observation of a dynamical process

Initial-state observation problem.

Given: The model $\mathcal{M}_f$ of a process
An I/O sequence $(\boldsymbol{U}, \boldsymbol{Y})(0 \dots \bar{k})$

Find: A state vector $\boldsymbol{x}_f^*(0)$ such that $\Psi_{y,f}(\boldsymbol{x}_f^*(0), \boldsymbol{U}(0 \dots \bar{k})) = \boldsymbol{Y}(0 \dots \bar{k})$

Final-state observation problem.

Given: The model $\mathcal{M}_f$ of a process
An I/O sequence $(\boldsymbol{U}, \boldsymbol{Y})(0 \dots \bar{k})$

Find: A state vector $\boldsymbol{x}_f^*(\bar{k})$ for which $\exists \boldsymbol{x}_0 \in \mathbb{R}^n$ such that
$\boldsymbol{x}_f^*(\bar{k}) := \psi_{x,f}(\boldsymbol{x}_0, \boldsymbol{U}(0 \dots \bar{k}-1))$ and $\Psi_{y,f}(\boldsymbol{x}_0, \boldsymbol{U}(0 \dots \bar{k})) = \boldsymbol{Y}(0 \dots \bar{k})$

A strong relationship exists between the initial and final observation problems. Indeed given an I/O sequence $(\boldsymbol{U}, \boldsymbol{Y})(0 \dots \bar{k})$, then the state $\boldsymbol{x}_0$ used in the description of the final-state observation problem is also a solution of the initial-state observation problem.

The distinction between these two problems is found in the literature in other terms. For example [88] refers to these two problems respectively as the *state-observation* and the *state-reconstruction*. In [57] these are described as *acausal* and *causal* observations (The initial-state observation is acausal since it uses future measurements ($\bar{k} > 0$) to determine the state at time 0). The final-state observation is described as an *attainability* problem in [67]. While the causality terminology clearly distinguishes the two problems, the terms "observation" and "reconstruction" are used without distinction in this thesis.

Noticeably, the solutions of the initial and final-state observation problems are *not* denoted with the superscript "o" which would indicate the true state of the system. This has multiple reasons. First, there is no statement made so far as to the observability of the system. Therefore, assuming that part of the state-space may not be observable, an infinity of solutions exists, of which only one is the true state. Second, even by assuming that the system is observable, the given I/O sequence may be too short to determine the (unique) true state of the system. (Compare this to a deadbeat observation scheme in which at least n successive measurements are needed to reconstruct the state [79].)

The most classical approach to state observation was designed by Luenberger [74] and reconstructs an estimate $\hat{\boldsymbol{x}}$ of the state which possesses the convergence property

$$\lim_{k\to\infty} \hat{\boldsymbol{x}}(k) = \boldsymbol{x}^\circ(k) \ . \tag{3.1}$$

For exact models (2.2)–(2.3), such an observer can be constructed if the Kalman criterion is verified, *i.e.*

$$\operatorname{rank} \begin{bmatrix} \boldsymbol{C}_f \\ \dots \\ \boldsymbol{C}_f \boldsymbol{A}_f^{n-1} \end{bmatrix} = n \ . \tag{3.2}$$

This criterion defines the *initial-state* observability and is a sufficient condition for the *final-state* observability. In the case of linear state-space systems, the two observability problems are equivalent if the system matrix $\boldsymbol{A}_f$ is regular, [60]. Kailath gives a simple example of this for the case of two autonomous first-order systems in parallel: $\boldsymbol{A}_f = \left[\begin{smallmatrix}1&1\\1&1\end{smallmatrix}\right]$, $\boldsymbol{C}_f = \left[\begin{smallmatrix}1&1\end{smallmatrix}\right]$. The system is final-state observable (reconstructible) since $x_1(k) = x_2(k) = y(k-1)$ but is not initial-state observable since one can only determine the sum $x_1(0) + x_2(0)$.

The convergence property (3.1) is lost as soon as the model or the input and output are not exactly known. To cope with these situations, robust observers are designed, of which the most popular and widespread example is the Kalman filter [61]. This method is based on the explicit consideration of disturbed I/O measurements supposed to be zero-mean and with known density functions (*e.g.* white noise). The Kalman filter is an *optimal observer* and hence computes a state $\hat{\boldsymbol{x}}$ which approximates the true state on average.

While this might be a satisfactory solution for many application cases, it is an insufficient solution for others. As the error

$$\boldsymbol{e} = \hat{\boldsymbol{x}} - \boldsymbol{x}$$

between the reconstructed and true states remains unknown, there is no *guarantee* on the location of the true state $\boldsymbol{x}^\circ$, based on its estimate $\hat{\boldsymbol{x}}$. For example, while it can be argued with an optimal observer that the process is *unlikely* in an undesirable area of the state-space (which may be a region of dangerous operating conditions), it cannot be *proven*.

Furthermore, a characterisation of the uncertainties in the I/O signals is to be used which is relevant to practical applications. It is in general difficult to obtain appropriate probabilistic assumptions for the measurement disturbances, whereas bounds on their amplitude are better known. This motivates the choice of pursuing a set-valued observation as described next.

3.2. Set-valued state observation

In the previous section, the state observation task was presented for real-valued input and output sequences $\boldsymbol{U}(0\ldots\bar{k})$ and $\boldsymbol{Y}(0\ldots\bar{k})$. Using the unknown-but-bounded measurement uncertainties presented in Section 2.3, leads to the consideration of set-valued input and output sequences $\mathcal{U}(0\ldots\bar{k})$ and $\mathcal{Y}(0\ldots\bar{k})$. Clearly a unique state sequence $\boldsymbol{X}^*$ cannot be found which fits all possible admissible I/O sequences $(\boldsymbol{U},\boldsymbol{Y}) \in (\mathcal{U},\mathcal{Y})$. Therefore, another kind of state observation is sought – a set-valued state observation – which describes a *set of states* which can occur given the sets of I/O and the model $\mathcal{M}_f$ of the system. The consideration of parameter uncertainties in the model further encourages the usage of such set-valued observer.

Definition 3.1 (State-Set Observer)
Given a model $\mathcal{M}_f$ and a sequence of I/O sets $(\mathcal{U},\mathcal{Y})(0\ldots\bar{k})$, an algorithm which reconstructs a sequence $\mathcal{X}_f(0\ldots\bar{k})$ of sets $\mathcal{X}_f(k) \subset \mathbb{R}^n$ is said to be a state-set observer.

As in the quantitative case, there is no unique realisation of the state-set observer. In fact Def. 3.1 is kept very open as to which algorithm is a state-set observer. Hence, more specific goals and properties are formulated in the following.

Problem descriptions. At first the initial-state and final-state observation problems are restated for the set-theoretical approach. This leads to the description of a set $\mathcal{X}_f^*$ containing the best possible (minimal) state reconstruction under consideration of all inputs and all outputs in the sequences of sets $\mathcal{U}$ and $\mathcal{Y}$. To preserve generality the problems are formulated for models subject to uncertainties (see the "$\exists \boldsymbol{P} \in \mathbb{P}^{k+1}$" quantifiers). The problem formulations for exact models are derived, as suggested in Section 2.2, by letting $\mathbb{P} = \{\mathbf{0}\}$ such that $\boldsymbol{P} = (\mathbf{0}, \ldots, \mathbf{0})$.

Initial-state set-observation problem.

Given: The model $\mathcal{M}_f$ of a process
A sequence of I/O sets $(\mathcal{U}, \mathcal{Y})(0 \ldots k)$

Find: The set of states consistent with an admissible I/O sequence:

$$\mathcal{X}_f^*(0 \,|\, 0 \ldots k) := \big\{\boldsymbol{x} \in \mathbb{R}^n \mid \exists \boldsymbol{U} \in \mathcal{U},\ \exists \boldsymbol{P} \in \mathbb{P}^{k+1}, \\ \text{s.t. } \Psi_{y,f,\boldsymbol{P}}(\boldsymbol{x}, \boldsymbol{U}(0 \ldots k)) \in \mathcal{Y}(0 \ldots k)\big\} \quad (3.3)$$

Final-state set-observation problem.

Given: The model $\mathcal{M}_f$ of a process
A sequence of I/O sets $(\mathcal{U}, \mathcal{Y})(0 \ldots k)$

Find: The set of states reached after an admissible I/O sequence:

$$\mathcal{X}_f^*(k \,|\, 0 \ldots k) := \big\{\boldsymbol{x} \in \mathbb{R}^n \mid \exists \boldsymbol{x}_0 \in \mathbb{R}^n,\ \exists \boldsymbol{U} \in \mathcal{U},\ \exists \boldsymbol{P} \in \mathbb{P}^{k+1} \\ \text{s.t. } \Psi_{y,f,\boldsymbol{P}}(\boldsymbol{x}_0, \boldsymbol{U}(0 \ldots k)) \in \mathcal{Y}(0 \ldots k) \\ \text{and } \boldsymbol{x} = \psi_{x,f,\boldsymbol{P}}(\boldsymbol{x}_0, \boldsymbol{U}(0 \ldots k-1))\big\} \quad (3.4)$$

In the above problems the notation $\mathcal{X}_f^*(0 \mid 0 \ldots k)$ is used to describe the initial-state set and is understood as "the set of states at time 0 given the measurement sequences between 0 and k." Similarly for the final-state set $\mathcal{X}_f^*(k \mid 0 \ldots k)$ which is "the set of states at time k given the measurement sequences between 0 and k."

The solution of both observation problems relate to one another as was already mentioned in Section 3.1. Indeed from any state $\boldsymbol{x}(0)$ in the minimal initial-state set, there exists an admissible I/O sequence leading to a state $\boldsymbol{x}(k)$ in the minimal final-state set. The reverse holds as well, [5]. This equivalence leads to the fact that the quantifier "$\exists \boldsymbol{x}_0 \in \mathbb{R}^n$" in the description of the final-state observation problem could be replaced with "$\exists \boldsymbol{x}_0 \in \mathcal{X}^*(0 \,|\, 0 \ldots k)$."

In the above problem descriptions, it is considered that absolutely no knowledge is known about the location of the system's state for $k < 0$. This is a worst-case assumption in which, before considering any measurement, the set of initial-states is assumed as the entire state-space. For both problems this is noted as an *a priori set*:

$$\mathcal{X}_f^*(0 \,|\, {-1}) = \mathbb{R}^n \,. \quad (3.5)$$

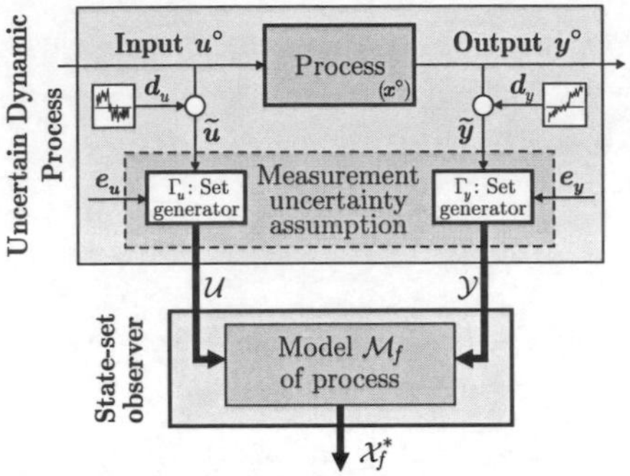

Figure 3.2.: State-set observation of a dynamical process subject to uncertainties.

Remark 3.2 – While this assumption is kept throughout the thesis, many similar results can be obtained by relaxing this condition. This happens if information about the state is given for the experiment. Using positive systems, for example, it is known that $\mathcal{X}_f^*(0\,|\,{-1}) = \mathbb{R}_+^n$. The two observation problems are then slightly reformulated such that the a priori set appears. The results of the observation problems then depend not only on the model and the sequence of I/O but also on this a priori set.

Remark 3.3 – For brevity of notation and when the observation horizon $0 \ldots k$ is clear from the context, the notation for the initial-state and final-state sets is shortened to $\mathcal{X}^*(0)$ and $\mathcal{X}^*(k)$ respectively.

Definition 3.2 (Minimal State-Set Observer) *Given a model $\mathcal{M}_f$ and a sequence of I/O sets $(\mathcal{U}, \mathcal{Y})(0 \ldots k)$, a state-set observer which reconstructs the set $\mathcal{X}_f^*(0\,|\,0 \ldots k)$ or $\mathcal{X}_f^*(k\,|\,0 \ldots k)$ is said to be a minimal state-set observer. The set $\mathcal{X}_f^*$ is referred to as the minimal state set.*

This definition describes the best possible state-set observer and is, therefore, used as a comparison for observation algorithms obtained in this thesis. It is noticed that the minimal state-set observer does not use any a priori assumption on the initial state set, as described by Eq. (3.5). Therefore it is minimal only with respect to the given sequences of I/O sets and the given model (as noted by the "f" subscript).

Theorem 2 (Minimal State-Set Contains the True State)
Given a complete model $\mathcal{M}_f$ and the I/O sets sequence $(\mathcal{U}, \mathcal{Y})(0 \ldots k)$, the true initial state $\boldsymbol{x}^\circ(0)$ (resp. true final state $\boldsymbol{x}^\circ(k)$) belongs to the minimal state set $\mathcal{X}_f^*(0\,|\,0 \ldots k)$ (resp.

minimal state set $\mathcal{X}_f^*(k\,|\,0\ldots k)$). Hence, the minimal state-set observer is guaranteed:

$$\boldsymbol{x}^\circ(0) \in \mathcal{X}_f^*(0\,|\,0\ldots k) \quad \textit{and} \quad \boldsymbol{x}^\circ(k) \in \mathcal{X}_f^*(k\,|\,0\ldots k)\ . \tag{3.6}$$

PROOF: By definition, the true state $\boldsymbol{x}^\circ(0)$ is consistent with the true I/O sequence $(\boldsymbol{U}^\circ, \boldsymbol{Y}^\circ)$ for the system behaviour. As the given model is complete, the same state $\boldsymbol{x}^\circ(0)$ is consistent with the true I/O sequence for the model $\mathcal{M}_f$. Since the true I/O sequence is admissible (Assumption III), *i.e.* $(\boldsymbol{U}^\circ, \boldsymbol{Y}^\circ) \in (\mathcal{U}, \mathcal{Y})$, and since the set $\mathcal{X}_f^*(0\,|\,0\ldots k)$ contains all states consistent with an admissible I/O sequence, it follows that $\boldsymbol{x}^\circ(0) \in \mathcal{X}_f^*(0\,|\,0\ldots k)$. A similar reasoning holds for $\boldsymbol{x}^\circ(k) \in \mathcal{X}(k)$. □

Definition 3.3 (Complete and Guaranteed Observation) *A state-set observer resulting in a set of states $\mathcal{X}_f$ is said to be* complete *iff $\mathcal{X}_f^* \subset \mathcal{X}_f$, and is* guaranteed *iff $\boldsymbol{x}^\circ \in \mathcal{X}_f$.*

From Theorem 2 it follows that a complete observation is also guaranteed (Fig. 3.3). While a non-minimal state set may contain the true state as well – in which case $\boldsymbol{x}^\circ \in (\mathcal{X}_f \cap \mathcal{X}_f^*)$ – this thesis focuses on the reconstruction of complete state sets in order to achieve guaranteed results.

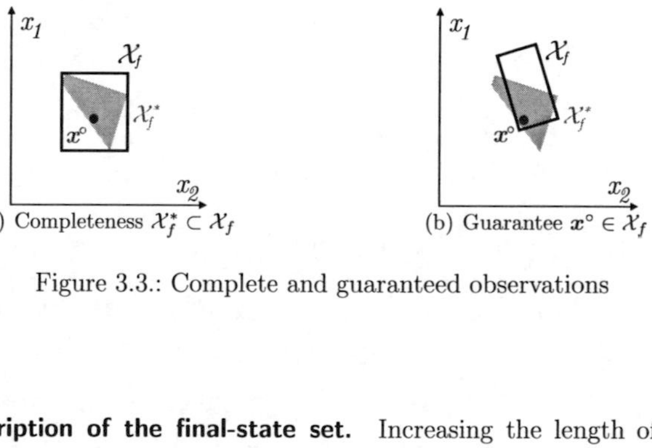

(a) Completeness $\mathcal{X}_f^* \subset \mathcal{X}_f$ (b) Guarantee $\boldsymbol{x}^\circ \in \mathcal{X}_f$

Figure 3.3.: Complete and guaranteed observations

Recursive description of the final-state set. Increasing the length of the sequences of I/O sets $(\mathcal{U}, \mathcal{Y})(0\ldots k)$ more information is obtained about the process. Hence, the shape and size of the sets $\mathcal{X}_f^*$ (both initial and final) depend on this measurement horizon. To emphasise this, the notation $\mathcal{X}_f^*(\cdot\mid 0\ldots k)$ was used to describe the initial-state and final-state sets.

Theorem 3 (Monotonicity of Minimal Initial-State Set)
Given a sequence of I/O sets $(\mathcal{U},\mathcal{Y})(0\dots k')$, the minimal initial state sets obtained for the entire I/O sequence and that obtained for a truncation of this sequence to the length $k<k'$ satisfy the relation:

$$\mathcal{X}_f^*(0\,|\,0\dots k) \supseteq \mathcal{X}_f^*(0\,|\,0\dots k'), \qquad k<k'. \tag{3.7}$$

PROOF: The description (3.3) of the initial-state observation problem for an exact model means that for all $\boldsymbol{x}_0 \in \mathcal{X}_f^*(0\,|\,0\dots k')$ there exists a sequence $\boldsymbol{U} \in \mathcal{U}(0\dots k')$, such that $\Psi_{y,f}(\boldsymbol{x}_0, \boldsymbol{U}(0\dots k')) \in \mathcal{Y}(0\dots k')$. The truncation of $\boldsymbol{U}$ and $\boldsymbol{Y}$ to the sequence of length $k<k'$, obviously verifies $\Psi_{y,f}(\boldsymbol{x}_0, \boldsymbol{U}(0\dots k)) \in \mathcal{Y}(0\dots k)$. Therefore $\boldsymbol{x}_0 \in \mathcal{X}_f^*(0\,|\,0\dots k)$, which proves Eq. (3.7) for the case of exact models. The extension for uncertain models is straightforward as the sequence of parameter $\boldsymbol{P}(0\dots k')$ may be truncated as well, leading to the same result. □

This result is important as it relates initial-state sets obtained for different lengths of the measurement sequences. It also implies that the set described in Eq. (3.3) can be solved *recursively* leading to an equivalent description of the minimal set of initial-states:

$$\mathcal{X}_f^*(0\,|\,0\dots k) := \big\{\boldsymbol{x} \in \mathcal{X}_f^*(0\,|\,0\dots k-1) \mid \exists \boldsymbol{U} \in \mathcal{U},\ \exists \boldsymbol{P} \in \mathbb{P}^{k+1}, \\ \text{s.t. } \Psi_{y,f,\boldsymbol{P}}(\boldsymbol{x}, \boldsymbol{U}(0\dots k)) \in \mathcal{Y}(0\dots k)\big\}\,. \tag{3.8}$$

An inclusion property such as that of Theorem 3 is not possible in the formulation of the final-state observation problem. Indeed, the set $\mathcal{X}_f^*(k\,|\,0\dots k)$ conceptually differs from $\mathcal{X}_f^*(k'\,|\,0\dots k')$, $k<k'$, since they describe states at different time instants. However, a recursive description of the minimal set of final-states is also possible, by decomposing the input-to-state and input-to-output operators used in Eq. (3.4), leading to the following proposition.

Proposition 3.1 (Recursive state-sets) *The sequence of sets $\mathcal{X}_f$ computed recursively*

$$\mathcal{X}_f(k) = \big\{\boldsymbol{x} \in \mathbb{R}^n \mid \boldsymbol{x} = \boldsymbol{A}_f(\boldsymbol{p})\hat{\boldsymbol{x}} + \boldsymbol{B}_f\hat{\boldsymbol{u}},\ \textit{and}\ (\boldsymbol{C}_f\boldsymbol{x} + \boldsymbol{D}_f\check{\boldsymbol{u}}) \in \mathcal{Y}(k), \\ \textit{with}\ \boldsymbol{p} \in \mathbb{P},\ \hat{\boldsymbol{x}} \in \mathcal{X}_f(k-1),\ \hat{\boldsymbol{u}} \in \mathcal{U}(k-1),\ \check{\boldsymbol{u}} \in \mathcal{U}(k)\big\} \tag{3.9}$$

and using the initial condition $\mathcal{X}_f(-1) = \mathbb{R}^n$ is complete, i.e. $\mathcal{X}_f(k) \supseteq \mathcal{X}_f^*(k|0\dots k)$.
For known models without direct feedthrough the equality $\mathcal{X}_f(k) = \mathcal{X}_f^*(k|0\dots k)$ holds, in which case the recursion is:

$$\mathcal{X}_f^*(k\,|\,0\dots k) = \big\{\boldsymbol{x} \in \mathbb{R}^n \mid \boldsymbol{x} = \boldsymbol{A}_f(\boldsymbol{p})\hat{\boldsymbol{x}} + \boldsymbol{B}_f\hat{\boldsymbol{u}},\ \textit{and}\ (\boldsymbol{C}_f\boldsymbol{x}) \in \mathcal{Y}(k), \\ \textit{with}\ \boldsymbol{p} \in \mathbb{P},\ \hat{\boldsymbol{x}} \in \mathcal{X}_f^*(k-1\,|\,0\dots k-1),\ \hat{\boldsymbol{u}} \in \mathcal{U}(k-1)\big\}\,. \tag{3.10}$$

■

PROOF: The first part of the proposition (general case leading to completeness) is proven by induction. The initial case is straightforward since $\mathcal{X}_f(-1) = \mathbb{R}^n$, as required from Eq. (3.5). Suppose the proposition holds true at time $k-1$, such that $\mathcal{X}_f(k-1) \supseteq \mathcal{X}^*(k-1\,|\,0\ldots k-1)$.

Using the definition of the final-state set-observation problem from Eq. (3.4), then $\forall \boldsymbol{x}(k) \in \mathcal{X}_f^*(k\,|\,0\ldots k)$ there exist an initial state $\boldsymbol{x}_0 \in \mathbb{R}^n$, a parameter sequence $\boldsymbol{P}(0\ldots k) \in \mathbb{P}^{k+1}$ and admissible input and output sequences $\boldsymbol{U} \in \mathcal{U}(0\ldots k)$ and $\boldsymbol{Y} \in \mathcal{Y}(0\ldots k)$ such that

$$\begin{aligned} &\Psi_{y,f,\boldsymbol{P}(0\ldots k)}(\boldsymbol{x}_0, \boldsymbol{U}(0\ldots k)) = \boldsymbol{Y}(0\ldots k) \\ \text{and } \boldsymbol{x}(k) =&\psi_{x,f,\boldsymbol{P}(0\ldots k-1)}(\boldsymbol{x}_0, \boldsymbol{U}(0\ldots k-1)) \ . \end{aligned}$$

By truncating the input, output and parameter sequences, there exists a state

$$\boldsymbol{x}(k-1) := \psi_{x,f,\boldsymbol{P}(0\ldots k-2)}(\boldsymbol{x}_0, \boldsymbol{U}(0\ldots k-2))$$

such that $\boldsymbol{x}(k-1) \in \mathcal{X}_f^*(k-1\,|\,0\ldots k-1)$ since the following also holds

$$\Psi_{y,f,\boldsymbol{P}(0\ldots k-1)}(\boldsymbol{x}_0, \boldsymbol{U}(0\ldots k-1)) = \boldsymbol{Y}(0\ldots k-1) \in \mathcal{Y}(0\ldots k-1) \ . \tag{3.11}$$

Therefrom the state $\boldsymbol{x}(k)$ may be described using $\boldsymbol{x}(k-1)$ as an initial state (instead of $\boldsymbol{x}_0$ as above) and using the same input, output and parameter values taken from the above sequences:

$$\begin{aligned} \boldsymbol{x}(k) =&\psi_{x,f,\boldsymbol{p}(k-1)}(\boldsymbol{x}(k-1), (\boldsymbol{u}(k-1))) \\ =&\boldsymbol{A}_f(\boldsymbol{p}(k-1))\boldsymbol{x}(k-1) + \boldsymbol{B}_f\boldsymbol{u}(k-1) \\ \text{and } &\psi_{y,f}(\boldsymbol{x}(k), (\boldsymbol{u}(k))) = \boldsymbol{C}_f\boldsymbol{x}(k) + \boldsymbol{D}_f\boldsymbol{u}(k) = \boldsymbol{y}(k) \ . \end{aligned} \tag{3.12}$$

Considering Eq. (3.9) for which

$$\begin{aligned} &\hat{\boldsymbol{u}} = \boldsymbol{u}(k-1) \in \mathcal{U}(k-1), \\ &\check{\boldsymbol{u}} = \boldsymbol{u}(k) \in \mathcal{U}(k), \\ &\hat{\boldsymbol{x}} = \boldsymbol{x}(k-1) \in \mathcal{X}_f(k-1) \quad \text{(induction assumption)}, \\ \text{and } &\boldsymbol{p} = \boldsymbol{p}(k-1) \in \mathbb{P}, \end{aligned}$$

it follows that $\boldsymbol{x}(k) \in \mathcal{X}_f(k)$. As this holds for all $\boldsymbol{x}(k) \in \mathcal{X}_f^*(k\,|\,0\ldots k)$, then $\mathcal{X}_f(k) \supseteq \mathcal{X}_f^*(k\,|\,0\ldots k)$ which finishes the proof by induction. The second part of the proposition is proven in [5]. □

Using Eq. (3.9), only the superset relation between $\mathcal{X}_f(k)$ and $\mathcal{X}_f^*(k\,|\,0\ldots k)$ is obtained in the general case because it cannot be assured that the input value $\boldsymbol{u}(k-1)$ used in Eq. (3.12) and the input value appearing in the sequence at time $k-1$ in Eq. (3.11) are the same. In other words, the input $\check{\boldsymbol{u}}$ used in computing $\mathcal{X}_f(k-1)$ and the input $\hat{\boldsymbol{u}}$ in computing $\mathcal{X}_f(k)$ cannot be assured to be identical. In interval arithmetic this is expressed as a problem of *broken dependency*, *i.e.* the desired relation $\check{\boldsymbol{u}} = \hat{\boldsymbol{u}}$ is lost as a result of the recursive and set-valued computation. This explains why the equality between $\mathcal{X}_f^*$ and $\mathcal{X}_f$ can only be claimed for the case of a known model with no direct feedthrough. In this case, Eq. (3.11) becomes

$$\Psi_{y,f,\boldsymbol{P}(0\ldots k-2)}(\boldsymbol{x}_0, \boldsymbol{U}(0\ldots k-2)) = \boldsymbol{Y}(0\ldots k-1)\ .$$

The recursive description given in Prop. 3.1 is the founding base for all algorithms presented in this thesis and in earlier publications dealing with set-valued observation (*e.g.* [103]). Such a decomposition is thinkable for any system described in state-space form, including nonlinear systems.

While the minimality of the resulting sets may be broken (hence the notation "$\mathcal{X}_f$" and not "$\mathcal{X}_f^*$"), the result is nevertheless *complete.* Thus, the most important results of this thesis are not altered from the recursive description of the state-sets, such that a guaranteed and complete diagnosis will be obtained in Chapter 4. (The breach of minimality is shown at an example in Appendix A.)

An illustration of the solutions to the initial-state and final-state set observation problems is found in Fig. 3.4. As can be seen from the left plot, the initial-state problem is a description of the possible states at time $k = 0$ and hence the described sets remain in the (x_1, x_2)-plane for $k = 0$. Having drawn the trajectory of the true state $\boldsymbol{x}^\circ$ (dashed-line), the figure illustrates the guarantee property (Theorem 2) and the monotonicity property (Theorem 3). From the right plot in the figure, it is seen that the final-state sets "follow" the trajectory of the true state hence describing in each time step the current possible location of the state. From the drawn trajectory of the true state $\boldsymbol{x}^\circ$, this also illustrates the guarantee property for the final-state set observation (Theorem 2).

3.3. Observation algorithms and properties

While two problems have been considered at the beginning of this chapter – the initial-state and the final-state observations – only one algorithm is presented: the reconstruction of a set of final-states. As described earlier these two problems are related but most of all – and from a practical point of view – the interest lies in describing the *current* state of a process (which is $\mathcal{X}^*(k)$) rather than in describing where a process has started from (which is $\mathcal{X}^*(0)$).

The following Algorithm I presents a very general form of state-set observation in the sense of Def. 3.1 and relies on the recursive description of the set observation problem given in Prop. 3.1. The algorithm's steps are graphically illustrated in Fig. 3.5 (for a model of second order with a single output). The algorithm uses operators denoted as $F_{f,p}$, $F_{f,m}$, $F_{\cap}$ and $F_{\approx}$ which are detailed afterwards.

Algorithm I (General State-Set Observation)

GIVEN:

- *The model $\mathcal{M}_f$ of the process*
- *The sequence of I/O sets $(\mathcal{U},\mathcal{Y})(0\dots\bar{k})$*
- *An a priori initial state set $\mathcal{X}(0\,|\,\text{-}1)\subseteq\mathbb{R}^n$*

INITIALISATION: $k := 0$

LOOP:

1. *Compute the predicted state set $\mathcal{X}_{f,p}$ (distinction $k=0$ and $k>0$ cases)*

$$\mathcal{X}_{f,p}(0) = \mathcal{X}(0\,|\,\text{-}1)$$
$$\text{and } \mathcal{X}_{f,p}(k) = F_{f,p}\big(\mathcal{X}_f(k-1),\mathcal{U}(k-1)\big) \tag{3.13}$$

2. *Compute the measured state set $\mathcal{X}_{f,m}$*

$$\mathcal{X}_{f,m}(k) = F_{f,m}\big(\mathcal{Y}(k),\mathcal{U}(k)\big) \tag{3.14}$$

3. *Compute the corrected state set $\mathcal{X}_{f,\cap}$*

$$\mathcal{X}_{f,\cap}(k) = F_{\cap}\big(\mathcal{X}_{f,p}(k),\mathcal{X}_{f,m}(k)\big) \tag{3.15}$$

4. *Compute the approximated set $\mathcal{X}_f(k)$*

$$\mathcal{X}_f(k) = F_{\approx}\big(\mathcal{X}_{f,\cap}(k)\big) \tag{3.16}$$

5. *If $k=\bar{k}$ stop, otherwise $k := k+1$ and go to Step 1.*

RESULT: *A sequence of final-state sets $\mathcal{X}_f(k)\equiv\mathcal{X}_f(k\,|\,0\dots k)$, $0\le k\le\bar{k}$.*

The algorithm forms a recursion in which $\mathcal{X}_f(k)$ is determined, for $k>0$, from the following

$$\mathcal{X}_f(k-1),\ \mathcal{U}(k-1),\ \mathcal{U}(k),\ \mathcal{Y}(k)\ \xrightarrow{\text{Alg. I}}\ \mathcal{X}_f(k)\ .$$

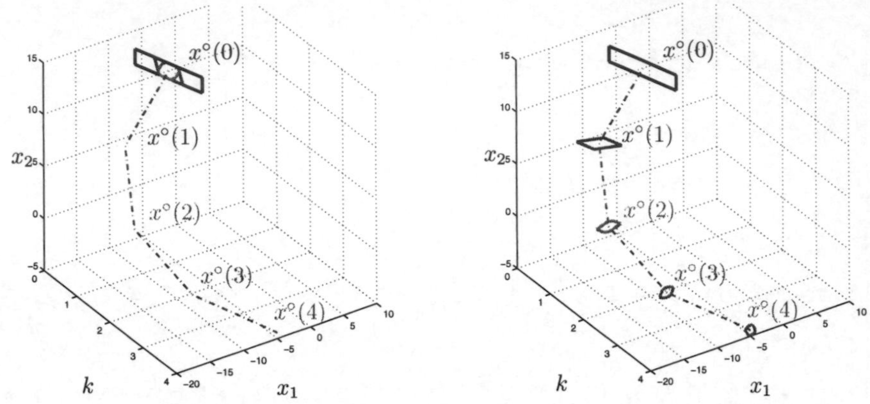

(a) Sequence of initial-state sets $\mathcal{X}_f^*(0 \mid 0 \dots \bar{k})$ (b) Sequence of final-state sets $\mathcal{X}_f^*(\bar{k} \mid 0 \dots \bar{k})$

Figure 3.4.: Illustration of the initial-state and final-state set-observations ($\bar{k} = 0, \dots, 4$). Legend: the solid lines show the border of the state-sets $\mathcal{X}^*$ and the dashed line represents the trajectory of the true state $\boldsymbol{x}^\circ$.

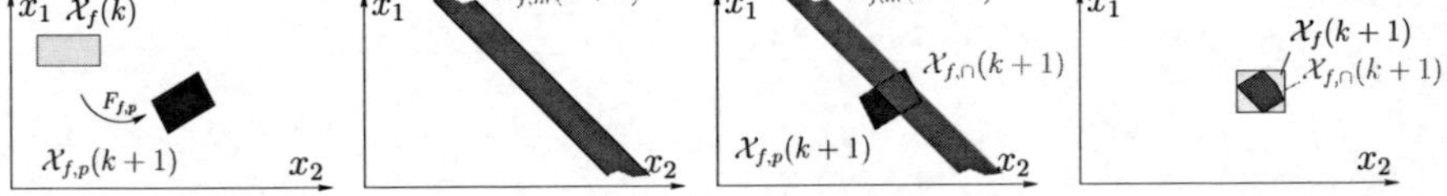

(a) Predicted set (b) Measured set (c) Corrected set (d) Approximated set

Figure 3.5.: Graphical illustration of the state-set observation steps. Example of a second order system, $\boldsymbol{x} = (x_1, x_2)^T$, with polytopic sets and axis-parallel (interval) approximation.

For $k = 0$ the recursion is understood as

$$\mathcal{X}(0\,|\,{-1}),\ \mathcal{U}(0),\ \mathcal{Y}(0) \quad \xrightarrow{Alg.\ \mathrm{I}} \quad \mathcal{X}_f(0)\ .$$

This decomposition is used in Chapter 4 to efficiently describe the diagnostic algorithm.

Both the choice of operators $F_\bullet$ ("$\bullet$" denotes any of the subscripts "f,p", "f,m", "$\cap$" or "$\approx$") and the choice of set formalism result in algorithms which possess specific properties such as minimality or completeness of the state-sets. As this description of the set observation algorithm is independent of the formalism used to represent the considered sets, it is general enough to encompass similar algorithms described in the past with subpavings [57, 92], with polytopes [120], with ellipsoids [104] or with zonotopes [30, 43, 44]. Despite this generality, the present thesis focuses on the use of polyhedric[1] sets, for which computational details are given in Section 3.4, including the case of uncertain models.

The operators $F_\bullet$ map sets to other sets:

$$F_{f,p}:\ 2^{\mathbb{R}^n} \times 2^{\mathbb{R}^m} \mapsto 2^{\mathbb{R}^n}, \qquad F_{f,m}:\ 2^{\mathbb{R}^r} \times 2^{\mathbb{R}^m} \mapsto 2^{\mathbb{R}^n},$$
$$F_{f,p}(\mathcal{X}_f,\ \mathcal{U}) = \mathcal{X}_{f,p} \qquad F_{f,m}(\mathcal{Y},\ \mathcal{U}) = \mathcal{X}_{f,m}$$

and

$$F_\cap:\ 2^{\mathbb{R}^n} \times 2^{\mathbb{R}^n} \mapsto 2^{\mathbb{R}^n}, \qquad F_\approx:\ 2^{\mathbb{R}^n} \mapsto 2^{\mathbb{R}^n}.$$
$$F_\cap(\mathcal{X}_{f,p},\ \mathcal{X}_{f,m}) = \mathcal{X}_{f,\cap} \qquad F_\approx(\mathcal{X}_{f,\cap}) = \mathcal{X}_f$$

Each of these operators have a *minimal formulation* $F_\bullet^*$ for which the most precise observation is achieved. These are described as:

$$\begin{aligned}
F_{f,p}^*(\mathcal{X},\mathcal{U}) :=& \{\boldsymbol{x} \mid \boldsymbol{x} = \psi_{x,f,\boldsymbol{p}}(\hat{\boldsymbol{x}}, (\hat{\boldsymbol{u}})),\ \hat{\boldsymbol{x}} \in \mathcal{X},\ \hat{\boldsymbol{u}} \in \mathcal{U},\ \boldsymbol{p} \in \mathbb{P}\} \\
=& \{\boldsymbol{x} \mid \boldsymbol{x} = \boldsymbol{A}_f(\boldsymbol{p})\hat{\boldsymbol{x}} + \boldsymbol{B}_f\hat{\boldsymbol{u}},\ \hat{\boldsymbol{x}} \in \mathcal{X},\ \hat{\boldsymbol{u}} \in \mathcal{U},\ \boldsymbol{p} \in \mathbb{P}\}, && (3.17) \\
F_{f,m}^*(\mathcal{Y},\mathcal{U}) :=& \{\boldsymbol{x} \mid \psi_{y,f,\boldsymbol{p}}(\boldsymbol{x}, (\check{\boldsymbol{u}})) \in \mathcal{Y},\ \check{\boldsymbol{u}} \in \mathcal{U},\ \boldsymbol{p} \in \mathbb{P}\} \\
=& \{\boldsymbol{x} \mid (\boldsymbol{C}_f\boldsymbol{x} + \boldsymbol{D}_f\hat{\boldsymbol{u}}) \in \mathcal{Y},\ \hat{\boldsymbol{u}} \in \mathcal{U}\}, && (3.18) \\
F_\cap^*(\mathcal{X}_1,\mathcal{X}_2) :=& \mathcal{X}_1 \cap \mathcal{X}_2\ , && (3.19) \\
F_\approx^*(\mathcal{X}) :=& \mathcal{X}\ \text{(identity operator)}\,. && (3.20)
\end{aligned}$$

Algorithm I is verbally described as the reconstruction of two distinct sets of possible states. In the first step, the information about the model's dynamics is used to predict its behaviour: using the set of states at time $k-1$ (noted $\mathcal{X}_f(k-1)$), a set of states *reachable*

[1] A polyhedron is a set inclosed by multiple inequality constraints, *cf.* [42, 122].

at time k (noted $\mathcal{X}_{f,p}(k)$) is constructed. In the second step, the relationship between measurements and states is used to construct a set of *measurable* states at time k (noted $\mathcal{X}_{f,m}(k)$). In the third step of the algorithm, the intersection of these two sets is formed such that only those states which are both reachable and measurable are kept.

The operator $F_\cap$ is introduced to preserve the generality of the algorithm for all formalisms. If ellipsoids are used then the intersection of two ellipsoids is not an ellipsoid. In this case a larger ellipsoid containing their intersection is described (which preserves completeness).

The operator $F_\approx$ is introduced to simplify the computation of the algorithm over time. For example the smallest interval set including the corrected set is used as overapproximation. Doing this in each final step of the loop keeps the complexity of the set description constant over time (see Section 3.4).

Different variants of Alg. I are now considered. At first an overapproximation operator is defined which is used for the complete algorithm (Alg. II), then a special case is considered for which the minimal state-sets are computed (Alg. III).

Definition 3.4 (Overapproximation Operator $\triangle$) *The operator $\triangle$ describes the overapproximation of a set $\mathcal{X} \subset \mathbb{R}^n$ given some predefined rules "$\boldsymbol{R}$":*

$$\triangle(\boldsymbol{R}, \mathcal{X}) \supseteq \mathcal{X} \ .$$

Algorithm II (Complete State-Set Observation)
The following implementation of Alg. I is considered with:

- *a model $\mathcal{M}_f(\mathbb{P})$, possibly with parameter uncertainties, Eqs.* (2.7)–(2.9)
- *the a priori set is $\mathcal{X}(0\,|\,{-1}) = \mathbb{R}^n$*
- *a polyhedric set formalism*
- *the operators $F_\bullet$ defined as:*

$$\begin{aligned} F_{f,p}(\mathcal{X},\mathcal{U}) &:= \triangle(\boldsymbol{M}_p, F^*_{f,p}(\mathcal{X},\mathcal{U})) \\ F_{f,m}(\mathcal{Y},\mathcal{U}) &:= F^*_{f,m}(\mathcal{Y},\mathcal{U}) \\ F_\cap(\mathcal{X}_1,\mathcal{X}_2) &:= \mathcal{X}_1 \cap \mathcal{X}_2 \\ F_\approx(\mathcal{X}) &:= \triangle(\boldsymbol{W},\mathcal{X}) \end{aligned}$$

with $F^*_\bullet$ as defined previously in Eqs. (3.17)–(3.20) *and where $\boldsymbol{M}_p$ and $\boldsymbol{W}$ are two rules defining the shape of the overapproximations of $F_{f,p}$ and $F_\approx$.*

The algorithm uses the Δ-operator to overapproximate the minimal operators. To this point, no assumptions are made on the choice of $\boldsymbol{M}_p$ or $\boldsymbol{W}$. A numerical example of this algorithm is found in Section 5.2, page 83.

Theorem 4 (Completeness of Algorithm II)
Algorithm II results in a sequence of complete final-state sets, i.e. $\mathcal{X}_f(k) \supseteq \mathcal{X}_f^*(k\,|\,0\ldots k)$.

PROOF: This theorem is proven by induction. The initial case is verified since the a priori set is chosen to be $\mathcal{X}(0\,|\,{-1}) = \mathbb{R}^n$ as in Eq. (3.5). This set can be represented as a polyhedron with no constraints. Suppose the proposition holds true at time $k-1$ such that the completeness $\mathcal{X}_f(k-1) \supseteq \mathcal{X}_f^*(k-1\,|\,0\ldots k-1)$ is known.

A recursion loop of Alg. II is expressed analytically using the $F_\bullet$ operators:

$$\begin{aligned}\mathcal{X}_f(k) &= F_\approx\left(F_\cap\left[F_{f,p}(\mathcal{X}_f(k-1),\mathcal{U}(k-1))\ ,\ F_{f,m}(\mathcal{Y}(k),\mathcal{U}(k))\right]\right)\\ &= F_\approx\left(F_{f,p}(\mathcal{X}_f(k-1),\mathcal{U}(k-1))\ \cap\ F_{f,m}^*(\mathcal{Y}(k),\mathcal{U}(k))\right)\ .\end{aligned}$$

Since $F_{f,p}(\cdot,\cdot) \supseteq F_{f,p}^*(\cdot,\cdot)$, the inclusion

$$\left(F_{f,p}(\cdot,\cdot) \cap F_{f,m}^*(\cdot,\cdot)\right) \supseteq \left(F_{f,p}^*(\cdot,\cdot) \cap F_{f,m}^*(\cdot,\cdot)\right)$$

holds, such that

$$\mathcal{X}_f(k) \supseteq F_\approx\left(F_{f,p}^*(\mathcal{X}_f(k-1),\mathcal{U}(k-1))\ \cap\ F_{f,m}^*(\mathcal{Y}(k),\mathcal{U}(k))\right)\ .$$

Furthermore, since $F_\approx(\cdot) \supseteq F_\approx^*(\cdot) = (\cdot)$ ($F_\approx^*$ is the identity), it follows

$$\mathcal{X}_f(k) \supseteq F_{f,p}^*(\mathcal{X}_f(k-1),\mathcal{U}(k-1))\ \cap\ F_{f,m}^*(\mathcal{Y}(k),\mathcal{U}(k))\ .$$

The above expression is recognised as the formulation of Eq. (3.9) using the minimal operators $F_{f,p}^*$ and $F_{f,m}^*$ as in Eqs. (3.17)–(3.18). As this formulation is proven in Prop 3.1 to lead to a complete set of final-states, hence

$$F_{f,p}^*(\mathcal{X}_f(k-1),\mathcal{U}(k-1))\ \cap\ F_{f,m}^*(\mathcal{Y}(k),\mathcal{U}(k)) \supseteq \mathcal{X}_f^*(k\mid 0\ldots k)$$

such that the sets computed by Alg. II are also complete since $\mathcal{X}_f(k) \supseteq \mathcal{X}_f^*(k\,|\,0\ldots k)$ holds. □

Algorithm III (Minimal State-Set Observation)
The following implementation of Alg. I is considered with:

- *a known model* $\mathcal{M}_f(\boldsymbol{p})$, *as in Eqs.* (2.2)–(2.3)
- *the model has no direct feedthrough (*i.e. $\boldsymbol{D}_f = \boldsymbol{0}$*)*
- *the a priori set is* $\mathcal{X}(0\,|\,{-1}) = \mathbb{R}^n$
- *a polyhedric set formalism*
- *the operators* $F_\bullet$ *defined as:*

$$\begin{aligned}
F_{f,p}(\mathcal{X},\mathcal{U}) &:= F^*_{f,p}(\mathcal{X},\mathcal{U}) = \{\boldsymbol{x} \in \mathbb{R}^n \mid \boldsymbol{x} = \boldsymbol{A}_f\hat{\boldsymbol{x}} + \boldsymbol{B}_f\hat{\boldsymbol{u}},\ \hat{\boldsymbol{x}} \in \mathcal{X},\ \hat{\boldsymbol{u}} \in \mathcal{U}\}\\
F_{f,m}(\mathcal{Y},\mathcal{U}) &:= F^*_{f,m}(\mathcal{Y},\varnothing) = \{\boldsymbol{x} \in \mathbb{R}^n \mid (\boldsymbol{C}_f\boldsymbol{x}) \in \mathcal{Y}\}\\
F_\cap(\mathcal{X}_1,\mathcal{X}_2) &:= F^*_\cap(\mathcal{X}_1,\mathcal{X}_2) = \mathcal{X}_1 \cap \mathcal{X}_2\\
F_\approx(\mathcal{X}) &:= F^*_\approx(\mathcal{X}) = \mathcal{X}\ .
\end{aligned}$$

Theorem 5 (Minimality of Algorithm III)
Algorithm III results in the sequence of minimal final-state sets, i.e. $\mathcal{X}_f(k) = \mathcal{X}^*_f(k\,|\,0\ldots k)$.

PROOF: This theorem is proven by induction, similarly as Theorem 4. The initial case is straightforward since $\mathcal{X}(0\,|\,{-1}) = \mathbb{R}^n$ is given and corresponds to Eq. (3.5).

The induction is based on the recursive description of the minimal state-sets from Eq. (3.10) in Prop. 3.1, adapted to the simpler case of exactly known models. As assumption to the inductive proof, suppose $\mathcal{X}_f(k-1) = \mathcal{X}^*_f(k-1\,|\,0\ldots k-1)$ and $\mathcal{X}_f(k-1)$ is a polyhedron. The measured sets $\mathcal{U}(k-1)$ and $\mathcal{Y}(k)$ are interval sets, hence the sets $F_{f,p}(\mathcal{X}_f(k-1),\mathcal{U}(k-1))$ and $F_{f,m}(\mathcal{Y}(k),\varnothing)$ can both be described exactly as polyhedra ($F_{f,p}$ is obtained by linear transformation and Minkowski-sum [122]). Since the intersection of two polyhedra is a polyhedron, the set

$$\mathcal{X}^*_f(k\,|\,0\ldots k) = F^*_{f,p}\big(\mathcal{X}^*_f(k-1),\mathcal{U}(k-1)\big)\ \cap\ F^*_{f,m}\big(\mathcal{Y}(k),\varnothing\big)$$

is the minimal state-set and is a polyhedron. This concludes the proof as this is what the algorithm computes:

$$\mathcal{X}_f(k) = F_{f,p}\big(\mathcal{X}_f(k-1),\mathcal{U}(k-1)\big)\ \cap\ F_{f,m}\big(\mathcal{Y}(k),\varnothing\big)$$

hence $\mathcal{X}_f(k) = \mathcal{X}^*_f(k\,|\,0\ldots k)$. □

Reading the above proof, one may wonder why the same result does not hold true if an uncertain model $\mathcal{M}_f(\mathbb{P})$ is considered in the algorithm. The issue lies in the realisation of the $F^*_{f,p}$ operator for uncertain models: it results in a non-convex set, [91]. The minimal prediction operator cannot be computed for any of the proposed formalisms. While the subpavings from [57] are not restricted to convex shapes, this method systematically results in an overapproximation, thus describing a complete prediction, but not minimal. This issue is described in more detail after Prop. 3.4 at hand of a numerical example.

A numerical example of minimal state-set observation is shown in Section 5.2, page 82.

Corollary 3.1 *According to Theorem 2, the two above Theorems 4 and 5 imply that the observation algorithms are guaranteed, hence*

$$\boldsymbol{x}^{\circ}(k) \in \mathcal{X}_f(k) \; .$$

Remark 3.4 – By not considering the output measurement, *i.e.* $F_{f,m}(\mathcal{Y},\mathcal{U}) := \mathbb{R}^n$, it is possible to realise a *set-valued simulation* of the model $\mathcal{M}_f$ based on the structure of Alg. I. It can then be shown that if $\boldsymbol{x}^{\circ}(0) \in \mathcal{X}(0\,|\,{-1})$ then $\boldsymbol{x}^{\circ}(k) \in \mathcal{X}_f(k)$, hence such a simulation is also guaranteed. In this case the recursion becomes

$$\mathcal{X}_f(k-1),\; \mathcal{U}(k-1) \quad \xrightarrow{set-simulation} \quad \mathcal{X}_f(k) \; .$$

3.4. Implementation using polyhedra

After showing the mathematical descriptions of polyhedric and interval sets, propositions are stated allowing the implementation of the set-observation algorithms for uncertain models (Alg. II) and for known models (Alg. III). The propositions are ordered according to their occurrence in the algorithm structure which is summarised in Table 3.1.

Observation step	Alg. II $\mathcal{M}_f(\boldsymbol{p})$	Alg. II $\mathcal{M}_f(\mathbb{P})$	Alg. III $\mathcal{M}_f(\boldsymbol{p})$, $\boldsymbol{D}_f = \boldsymbol{0}$
1. Predicted set	Prop. 3.3	Prop. 3.4	Prop. 3.2
2. Measured set	Prop. 3.5	Prop. 3.5	Prop. 3.5
3. Corrected set	Prop. 3.6	Prop. 3.6	Prop. 3.6
4. Approximated set	Prop. 3.8	Prop. 3.8	(Prop. 3.7)†

Table 3.1.: Classification of propositions for implementation of set-observers.

(† This step is not needed theoretically but maintains the numerical tractability over time)

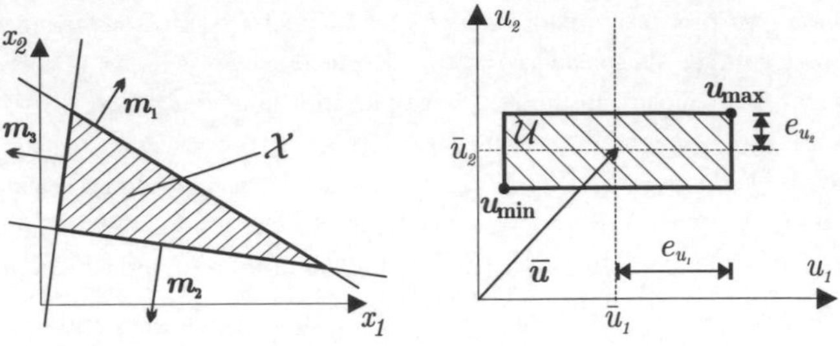

(a) State set $\mathcal{X}$ as polyhedric set.

(b) Input set $\mathcal{U}$ as interval set. (In general $\bar{\boldsymbol{u}} = \tilde{\boldsymbol{u}}$)

Figure 3.6.: Formalism used for state sets and for I/O sets.

The presented implementation was realised in MATLAB™ and is briefly presented in Appendix B. Numerical results are shown in Chapter 5.

Additionally to the properties stated in Section 3.3, the implementation should possess interesting numerical properties leading to tractable computations. This aspect is discussed in more detail at the end of this section.

Choice of set-valued formalism

It is chosen to work with polyhedric sets since these allow the reconstruction of the minimal set of final states (*cf.* Theorem 5) and offer a good compromise between their mathematical complexity and geometrical flexibility. Therefore, the implementation considers polyhedric state sets and interval input and output sets. Examples of such sets are seen in Fig. 3.6.

Polyhedric sets are used to represent the state sets as they offer a flexible representation of (convex) geometrical forms, [42, 122]. A polyhedron is defined by s inequalities,[2] each describing one of its faces as a hyperspace: (see Fig. 3.6(a))

$$\begin{aligned} \mathcal{X}(k) &= \{\boldsymbol{x} \in \mathbb{R}^n \mid \boldsymbol{m}_i^T \boldsymbol{x} \le q_i,\ i = 1, \dots, s\} \\ &= \{\boldsymbol{x} \in \mathbb{R}^n \mid \boldsymbol{M}\boldsymbol{x} \le \boldsymbol{q}\},\ \text{with } \boldsymbol{M} = \begin{bmatrix} \boldsymbol{m}_1^T \\ \cdots \\ \boldsymbol{m}_s^T \end{bmatrix} \text{ and } \boldsymbol{p} = (q_1, \dots, q_s)^T\ . \end{aligned} \tag{3.21}$$

The rows of $\boldsymbol{M}$ describe the faces' orientations and the components of $\boldsymbol{q}$ the according faces's position. For a given orientation $\boldsymbol{m}_i^T$, increasing the value of q_i increases the size of

[2]In general polyhedra may also contain equality constraints leading to a slightly different formalism not used here. There also exist open-polyhedra defined using strict inequalities "<".

the polyhedron by moving the corresponding boundary away from the polyhedron (in the direction of its orientation vector $\boldsymbol{m}_i^T$).

Set-valued observation problems often come down to solving optimisation problems, even in the simplest interval case *e.g.* [94]. One large advantage of using polyhedra is that the encountered optimisations are *linear programs*, far easier to solve than generic optimisations [21, 22].

While $\boldsymbol{M}$ and $\boldsymbol{q}$ define a unique polyhedron $\mathcal{X}$, this same set may be represented in an infinity of different ways, [122]. Indeed the set is invariant by scaling of the constraints and by adding or removing redundant constraints (see Fig. 3.7). There exist a unique representation of the set $\mathcal{X}$ in which no constraint is redundant and all orientation vectors are normalised ($\|\boldsymbol{m}_i\| = 1,\ \forall i$). This is the *normalised minimal representation* of the polyhedron [20, 108].

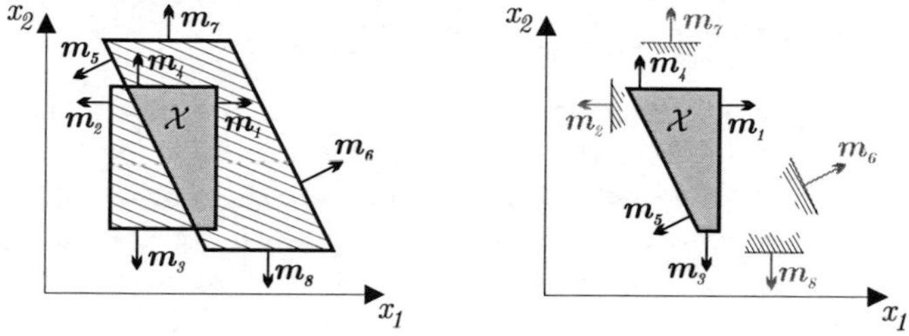

(a) With redundant constraints (b) Minimal representation

Figure 3.7.: Polytopic sets with and without redundant constraints, $\mathcal{X} = \{(\boldsymbol{m}_i, q_i)\}_{i=1,\ldots,8} = \{(\boldsymbol{m}_i, q_i)\}_{i=1,3,4,5}$.

Interval sets (or hypercubes) are a special case of polyhedra which are bounded and which parallel faces are oriented by the base vectors of the space (see Fig. 3.6(b)). Therefore, an interval set is represented as

$$\mathcal{U}(k) = \left\{\boldsymbol{u} \in \mathbb{R}^m \mid \left[\begin{smallmatrix} \boldsymbol{I}_m \\ -\boldsymbol{I}_m \end{smallmatrix}\right]\boldsymbol{u} \leq \left(\begin{smallmatrix} \boldsymbol{u}_{\max} \\ -\boldsymbol{u}_{\min} \end{smallmatrix}\right)\right\}. \tag{3.22}$$

A simpler yet equivalent formulation exists in which each component of the space is delimited by a minimum and maximum bound:

$$\begin{aligned} \mathcal{U}(k) &= \{\boldsymbol{u} \in \mathbb{R}^m \mid u_{i,\min} \leq u_i \leq u_{i,\max},\ i = 1, \ldots, m\} \\ &= \{\boldsymbol{u} \in \mathbb{R}^m \mid \boldsymbol{u}_{\min} \leq \boldsymbol{u} \leq \boldsymbol{u}_{\max}\} \end{aligned} \tag{3.23}$$

with $\boldsymbol{u}_{\min} := (u_{1,\min}, \ldots, u_{m,\min})^T$, $\boldsymbol{u}_{\max} := (u_{1,\max}, \ldots, u_{m,\max})^T$. As drawn in Fig. 3.6(b), the middle of an interval set is obtained with the mid-operator which computes

$$\operatorname{mid}(\mathcal{U}) := \frac{\boldsymbol{u}_{\max} - \boldsymbol{u}_{\min}}{2} = \bar{\boldsymbol{u}} \; .$$

Similarly the output set is described by

$$\mathcal{Y}(k) = \{\boldsymbol{y} \in \mathbb{R}^r \mid \boldsymbol{y}_{\min} \leq \boldsymbol{y} \leq \boldsymbol{y}_{\max}\} \quad \text{and} \quad \operatorname{mid}(\mathcal{Y}) = \bar{\boldsymbol{y}} \; . \tag{3.24}$$

This chosen formalism for the input and output sets allows to easily represent unknown-but-bounded measurement errors, as described in Section 2.3.

Remark 3.5 – Bounded polyhedra, or *polytopes*, are represented using either the above hyperspace notation (known as an $\mathcal{H}$-polytope) or as a list of its vertices (a $\mathcal{V}$-polytope), [42, 122]. In the latter case, the polytope is described as

$$\mathcal{X}(k) = \operatorname{conv}\left(\{\boldsymbol{v}_1,\ \boldsymbol{v}_2,\ \ldots\}\right) ,$$

where “conv” denotes the computation of the convex hull of a set of points in space. As there exist problems which are easily solved using one notation but become intractable in the other, the choice of the notation is usually important. However, because the set-valued observation algorithms require the description of unbounded sets (starting with the unbounded a priori set $\mathcal{X}(0 \mid -1)$) the hyperspace notation is used throughout this thesis – with the exception of the proof of Prop. 3.9.

In general, the boundedness of a polyhedron can only be verified using a numerical criterion, see *e.g.* [64]. For the special case of a *symmetric polyhedron*, which has the form

$$\mathcal{X} = \{\boldsymbol{x} \in \mathbb{R}^n \mid [\begin{smallmatrix} \boldsymbol{M} \\ -\boldsymbol{M} \end{smallmatrix}]\boldsymbol{x} \leq (\begin{smallmatrix} \boldsymbol{q} \\ \boldsymbol{q} \end{smallmatrix})\} , \tag{3.25}$$

the boundedness is assured under the condition: $\operatorname{rank}(\boldsymbol{M}) = n$, [72].

Computing the predicted set

Lemma 3.1 (Linear transformation of polyhedron) *Given a polyhedron $\mathcal{X}$ defined by Eq.* (3.21) *and a regular matrix $\boldsymbol{A}_f \in \mathbb{R}^{n \times n}$, the set $\mathcal{X}'$ resulting from the linear transformation $(\boldsymbol{A}_f \boldsymbol{x})$, $\forall \boldsymbol{x} \in \mathcal{X}$, is a polyhedron described by*

$$\mathcal{X}' = \{\boldsymbol{x} = \boldsymbol{A}_f \hat{\boldsymbol{x}}, \hat{\boldsymbol{x}} \in \mathcal{X}\} = \{\boldsymbol{x} \in \mathbb{R}^n \mid \boldsymbol{M}\boldsymbol{A}_f^{-1}\boldsymbol{x} \leq \boldsymbol{q}\} \; .$$

This transformation is noted as $\mathcal{X}' = \boldsymbol{A}_f \mathcal{X}$. ■

PROOF: Since $\boldsymbol{A}_f$ is regular, it can be reformulated that for all $\boldsymbol{x} \in \mathcal{X}'$, $\exists \hat{\boldsymbol{x}} \in \mathcal{X}$ such that $\hat{\boldsymbol{x}} = \boldsymbol{A}_f^{-1}\boldsymbol{x}$. Since $\boldsymbol{M}\hat{\boldsymbol{x}} \leq \boldsymbol{q}$ it results that $\boldsymbol{M}\boldsymbol{A}_f^{-1}\boldsymbol{x} \leq \boldsymbol{q}$. □

Proposition 3.2 (Prediction for exact model) *Given a polyhedron $\mathcal{X}$ defined by Eq. (3.21), an input set $\mathcal{U}$ defined by Eq. (3.22), the matrices $\boldsymbol{A}_f$, $\boldsymbol{B}_f$ describing an exact model $\mathcal{M}_f(\boldsymbol{p})$ as in Eqs. (2.2)–(2.3), then the prediction $\mathcal{X}^*_{f,p}$ is obtained as*

$$\mathcal{X}^*_{f,p} = F^*_{f,p}(\mathcal{X},\mathcal{U}) = \{\boldsymbol{x} \in \mathbb{R}^n \mid \boldsymbol{P}\boldsymbol{x} \leq \boldsymbol{g}\}$$

where $\boldsymbol{P}$ and $\boldsymbol{g}$ are obtained from a Fourier-Motzkin elimination[3] *of the last m dimensions of the set*

$$\mathcal{P} = \left\{ (\boldsymbol{x},\boldsymbol{u}) \in \mathbb{R}^n \times \mathbb{R}^m \mid \begin{bmatrix} \boldsymbol{M}\boldsymbol{A}_f^{-1} & -\boldsymbol{M}\boldsymbol{A}_f^{-1}\boldsymbol{B}_f \\ \boldsymbol{0} & \boldsymbol{I}_m \\ \boldsymbol{0} & -\boldsymbol{I}_m \end{bmatrix} \begin{pmatrix} \boldsymbol{x} \\ \boldsymbol{u} \end{pmatrix} \leq \begin{pmatrix} \boldsymbol{q} \\ \boldsymbol{u}_{\max} \\ -\boldsymbol{u}_{\min} \end{pmatrix} \right\} . \qquad \blacksquare$$

PROOF: Lets recall the definition of the minimal operator $F^*_{f,p}$, described in Eq. (3.17), for an exact model

$$F^*_{f,p}(\mathcal{X},\mathcal{U}) = \{\boldsymbol{x} \in \mathbb{R}^n \mid \boldsymbol{x} = \boldsymbol{A}_f\hat{\boldsymbol{x}} + \boldsymbol{B}_f\boldsymbol{u},\ \hat{\boldsymbol{x}} \in \mathcal{X},\ \boldsymbol{u} \in \mathcal{U}\} . \tag{3.26}$$

This set is the projection of

$$\mathcal{P}' = \{(\boldsymbol{x},\hat{\boldsymbol{x}},\boldsymbol{u}) \in \mathbb{R}^n \times \mathbb{R}^n \times \mathbb{R}^m \mid \boldsymbol{x} = \boldsymbol{A}_f\hat{\boldsymbol{x}} + \boldsymbol{B}_f\boldsymbol{u},\ \hat{\boldsymbol{x}} \in \mathcal{X},\ \boldsymbol{u} \in \mathcal{U}\}$$

on its first coordinate $\boldsymbol{x}$. The projection problem is simplified using the fact that $\boldsymbol{A}_f\hat{\boldsymbol{x}} = (\boldsymbol{x} - \boldsymbol{B}_f\boldsymbol{u}) \in \boldsymbol{A}_f\mathcal{X}$ and hence, $\boldsymbol{M}\boldsymbol{A}_f^{-1}(\boldsymbol{x} - \boldsymbol{B}_f\boldsymbol{u}) \leq \boldsymbol{q}$ (Lemma 3.1). Using the polyhedric description of the input set $\mathcal{U}$ from Eq. (3.22), the above set (3.26) is therefore the projection of

$$\mathcal{P} = \{(\boldsymbol{x},\boldsymbol{u}) \in \mathbb{R}^n \times \mathbb{R}^m \mid (\boldsymbol{x} - \boldsymbol{B}_f\boldsymbol{u}) \in \boldsymbol{A}_f\mathcal{X} \text{ and } \boldsymbol{u} \in \mathcal{U}\}$$
$$\mathcal{P} = \{(\boldsymbol{x},\boldsymbol{u}) \in \mathbb{R}^n \times \mathbb{R}^m \mid \boldsymbol{M}\boldsymbol{A}_f^{-1}(\boldsymbol{x} - \boldsymbol{B}_f\boldsymbol{u}) \leq \boldsymbol{q} \text{ and } \left[\begin{smallmatrix} \boldsymbol{I}_m \\ -\boldsymbol{I}_m \end{smallmatrix}\right]\boldsymbol{u} \leq \left(\begin{smallmatrix} \boldsymbol{u}_{\max} \\ -\boldsymbol{u}_{\min} \end{smallmatrix}\right)\} \qquad \square$$

on its first coordinate $\boldsymbol{x}$. The above set $\mathcal{P}$ is equal to the formulation found in the proposition by expressing all inequalities in terms of the single vector variable $\left(\begin{smallmatrix} \boldsymbol{x} \\ \boldsymbol{u} \end{smallmatrix}\right)$.

The projection achieved in Prop. 3.2 results in a polyhedron $\mathcal{X}^*_{f,p}$ which contains at least as many constraints as $\mathcal{X}$ (and usually many redundant constraints). In [5], it is shown that the orientation of s constraints are directly obtained from the linear transformation of $\mathcal{X}$

[3] The Fourier-Motzkin elimination is not described here but algorithms are found in the literature, *e.g.* [63, 101]. It is the counterpart for systems of inequalities to the well-known Gaussian elimination used for systems of equalities.

for the system matrix $\boldsymbol{A}_f$. Hence, $\mathcal{X}^*_{f,p}$ has s constraints oriented by $\boldsymbol{M}\boldsymbol{A}_f^{-1}$ and additional constraints which orientations result from the Fourier-Motzkin elimination.

By removing these additional constraints, a larger set $\mathcal{X}_{f,p} \supset \mathcal{X}^*_{f,p}$ is obtained. As shown in the next proposition, such a predicted set $\mathcal{X}_{f,p}$ can be computed without doing the Fourier-Motzkin elimination at all. This causes a loss of accuracy in the prediction step but improves the overall computational effort since the resulting set is mathematically simpler (*i.e.* has less constraints, see Fig. 3.8). Furthermore, the idea behind this simplification is used in Prop. 3.4 for the prediction of models with uncertain parameters.

Proposition 3.3 (Simplified prediction for exact model) *Given a polyhedron $\mathcal{X}$ defined by Eq.* (3.21), *an input set $\mathcal{U}$ defined by Eq.* (3.22), *the matrices $\boldsymbol{A}_f$ and $\boldsymbol{B}_f$ describing an exact model $\mathcal{M}_f(\boldsymbol{p})$ as in Eqs.* (2.2)–(2.3), *then a simplified prediction $\mathcal{X}_{f,p}$ is given by*

$$\mathcal{X}_{f,p} = \triangle\big(\boldsymbol{M}\boldsymbol{A}_f^{-1}, F^*_{f,p}(\mathcal{X},\mathcal{U})\big) = \{\boldsymbol{x} \in \mathbb{R}^n \mid \boldsymbol{M}\boldsymbol{A}_f^{-1}\boldsymbol{x} \le \boldsymbol{h}\}$$

with $\boldsymbol{h} = (h_1, \dots, h_s)^T$ obtained as

$$h_i := q_i + \max_{\boldsymbol{u}\in\mathcal{U}}(\boldsymbol{m}_i^T \boldsymbol{A}_f^{-1}\boldsymbol{B}_f\boldsymbol{u}), \quad i = 1, \dots, s \ . \qquad \blacksquare$$

PROOF: It needs to be shown that the proposed polyhedric description of $\mathcal{X}_{f,p}$ overapproximates the $\mathcal{X}^*_{f,p}$ computed in Prop. 3.2. For all $\boldsymbol{x} \in \mathcal{X}^*_{f,p} = F^*_{f,p}(\mathcal{X},\mathcal{U})$, $\exists\hat{\boldsymbol{x}} \in \mathcal{X}$ and $\exists\boldsymbol{u} \in \mathcal{U}$ such that $\boldsymbol{x} = \boldsymbol{A}_f\hat{\boldsymbol{x}} + \boldsymbol{B}_f\boldsymbol{u}$, or $(\boldsymbol{x} - \boldsymbol{B}_f\boldsymbol{u}) \in \boldsymbol{A}_f\mathcal{X}$. Therefore, using Lemma 3.1, $\boldsymbol{M}\boldsymbol{A}_f^{-1}(\boldsymbol{x} - \boldsymbol{B}_f\boldsymbol{u}) \le \boldsymbol{q}$ and

$$\boldsymbol{M}\boldsymbol{A}_f^{-1}\boldsymbol{x} \le (\boldsymbol{q} + \boldsymbol{M}\boldsymbol{A}_f^{-1}\boldsymbol{B}_f\boldsymbol{u}), \quad \boldsymbol{u} \in \mathcal{U} \ . \tag{3.27}$$

This vectorial inequality is decomposed into the following scalar inequalities

$$\boldsymbol{m}_i^T\boldsymbol{A}_f^{-1}\boldsymbol{x} \le (q_i + \boldsymbol{m}_i^T\boldsymbol{A}_f^{-1}\boldsymbol{B}_f\boldsymbol{u}), \quad i = 1, \dots, s, \ \boldsymbol{u} \in \mathcal{U} \ .$$

Let $h'_i := \max_{\boldsymbol{u}\in\mathcal{U}}(\boldsymbol{m}_i^T\boldsymbol{A}_f^{-1}\boldsymbol{B}_f\boldsymbol{u})$ such that the s inequalities are overapproximated using the upper bound

$$\boldsymbol{m}_i^T\boldsymbol{A}_f^{-1}\boldsymbol{x} \le (q_i + h'_i) =: h_i \ .$$

Therefore, $\forall\boldsymbol{x} \in \mathcal{X}^*_{f,p}$, the vectorial inequality $\boldsymbol{M}\boldsymbol{A}_f^{-1}\boldsymbol{x} \le \boldsymbol{h}$ holds and $\boldsymbol{x} \in \mathcal{X}_{f,p}$.

The breach of minimality, leading to $\mathcal{X}_{f,p} \ne \mathcal{X}^*_{f,p}$, is due to the computation of s distinct optimisations in the input variable $\boldsymbol{u}$. While an optimum input is found *for each constraint* $\boldsymbol{m}_i^T\boldsymbol{A}_f^{-1}\boldsymbol{B}_f$, there may not exist a single input solution of *all* s constraints. □

Corollary 3.2 *Props. 3.2 and 3.3 are equivalent if no input uncertainties are considered ($\mathcal{U} = \{\boldsymbol{u}\}$) or if the model is autonomous ($\boldsymbol{B}_f = \boldsymbol{0}$).*

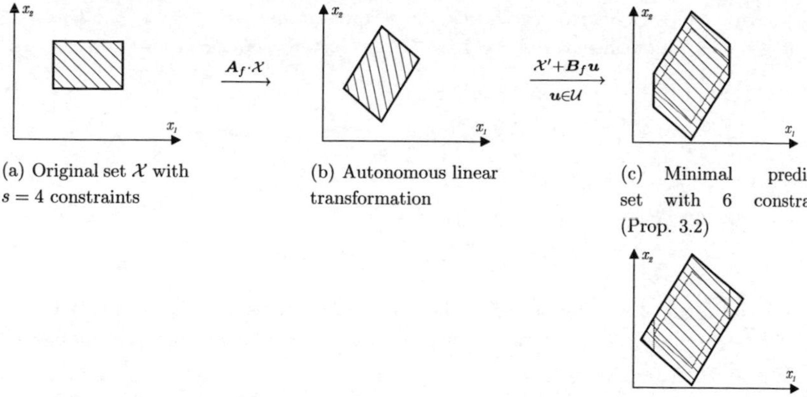

(a) Original set $\mathcal{X}$ with $s = 4$ constraints

(b) Autonomous linear transformation

(c) Minimal predicted set with 6 constraints (Prop. 3.2)

(d) Simplified predicted set with 4 constraints (Prop. 3.3)

Figure 3.8.: Minimal and simplified prediction of a two-dimensional state set.

Proposition 3.4 (Simplified prediction for uncertain model) *Given a polyhedron $\mathcal{X}$ defined by Eq. (3.21), an input set $\mathcal{U}$ defined by Eq. (3.22), an uncertain model $\mathcal{M}_f(\mathbb{P})$ as in Eqs. (2.7)–(2.9) and (2.10) for which $\boldsymbol{A}_{f,0}$ is regular. A simplified prediction $\mathcal{X}_{f,p}$ is obtained as*

$$\mathcal{X}_{f,p} = \triangle\big(\boldsymbol{M}\boldsymbol{A}_{f,0}^{-1}, F_{f,p}^*(\mathcal{X},\mathcal{U})\big) = \{\boldsymbol{x} \in \mathbb{R}^n \mid \boldsymbol{M}\boldsymbol{A}_{f,0}^{-1}\boldsymbol{x} \leq \boldsymbol{h}\}$$

with $\boldsymbol{h} = (h_1, \ldots, h_s)^T$ obtained as

$$h_i := q_i + \max_{\boldsymbol{u}\in\mathcal{U}}(\boldsymbol{m}_i^T \boldsymbol{A}_{f,0}^{-1}\boldsymbol{B}_f\boldsymbol{u}) + \sum_{j=1}^{N_p} \max\Big[p_j^{min}\min_{\boldsymbol{x}\in\mathcal{X}}(\boldsymbol{m}_i^T\boldsymbol{A}_{f,0}^{-1}\boldsymbol{A}_{f,j}\boldsymbol{x}),\ p_j^{min}\max_{\boldsymbol{x}\in\mathcal{X}}(\boldsymbol{m}_i^T\boldsymbol{A}_{f,0}^{-1}\boldsymbol{A}_{f,j}\boldsymbol{x}),$$
$$p_j^{max}\min_{\boldsymbol{x}\in\mathcal{X}}(\boldsymbol{m}_i^T\boldsymbol{A}_{f,0}^{-1}\boldsymbol{A}_{f,j}\boldsymbol{x}),\ p_j^{max}\max_{\boldsymbol{x}\in\mathcal{X}}(\boldsymbol{m}_i^T\boldsymbol{A}_{f,0}^{-1}\boldsymbol{A}_{f,j}\boldsymbol{x})\Big]\ . \tag{3.28}$$

■

PROOF: At first lets point out that Props. 3.3 and 3.4 are identical if the latter is applied to a known model. Indeed, in this case $\mathbb{P} = \{\boldsymbol{0}\}$ holds (*i.e.* $p_j^{min} = p_j^{max} = 0$), such that the final sum vanishes. Furthermore $\boldsymbol{A}_{f,0}$ (of the uncertain model) and $\boldsymbol{A}_f$ (of the known model) are equal.

The proof is similar to that of Prop. 3.3 and it has to be proven that the computed set $\mathcal{X}_{f,p}$ includes the minimal $\mathcal{X}_{f,p}^* = F_{f,p}^*(\mathcal{X},\mathcal{U})$. For all $\boldsymbol{x} \in \mathcal{X}_{f,p}^*$, $\exists \boldsymbol{u} \in \mathcal{U}$, $\exists \hat{\boldsymbol{x}} \in \mathcal{X}$ and $\exists \boldsymbol{p} \in \mathbb{P}$ such that $\boldsymbol{x} = \boldsymbol{A}_f(\boldsymbol{p})\hat{\boldsymbol{x}} + \boldsymbol{B}_f\boldsymbol{u}$. Given the constraints orientated by $\boldsymbol{M}\boldsymbol{A}_{f,0}^{-1}$, an overapproximation of $\mathcal{X}_{f,p}^*$

is obtained by finding the vector $\boldsymbol{h}$ such that the vectorial inequality $\boldsymbol{M}\boldsymbol{A}_{f,0}^{-1}\boldsymbol{x} \leq \boldsymbol{h}$ holds. By considering the individual inequalities, $i = 1, \dots, s$, the following statements are made

$$\begin{aligned}
\boldsymbol{m}_i^T \boldsymbol{A}_{f,0}^{-1}\boldsymbol{x} &= \boldsymbol{m}_i^T \boldsymbol{A}_{f,0}^{-1}(\boldsymbol{A}_f(\boldsymbol{p})\hat{\boldsymbol{x}} + \boldsymbol{B}_f\boldsymbol{u}) \\
&= \boldsymbol{m}_i^T \boldsymbol{A}_{f,0}^{-1}\boldsymbol{A}_f(\boldsymbol{p})\hat{\boldsymbol{x}} + \boldsymbol{m}_i^T \boldsymbol{A}_{f,0}^{-1}\boldsymbol{B}_f\boldsymbol{u} \\
&\leq \max_{\substack{\boldsymbol{p}\in\mathbb{P}\\ \hat{\boldsymbol{x}}\in\mathcal{X}\\ \boldsymbol{u}\in\mathcal{U}}} (\boldsymbol{m}_i^T \boldsymbol{A}_{f,0}^{-1}\boldsymbol{A}_f(\boldsymbol{p})\hat{\boldsymbol{x}} + \boldsymbol{m}_i^T \boldsymbol{A}_{f,0}^{-1}\boldsymbol{B}_f\boldsymbol{u}) \\
&\leq \max_{\substack{\boldsymbol{p}\in\mathbb{P}\\ \hat{\boldsymbol{x}}\in\mathcal{X}}} (\boldsymbol{m}_i^T \boldsymbol{A}_{f,0}^{-1}\boldsymbol{A}_f(\boldsymbol{p})\hat{\boldsymbol{x}}) + \max_{\boldsymbol{u}\in\mathcal{U}} (\boldsymbol{m}_i^T \boldsymbol{A}_{f,0}^{-1}\boldsymbol{B}_f\boldsymbol{u}) \, .
\end{aligned} \tag{3.29}$$

While the second maximum may be easily computed ($\mathcal{U}$ in an interval set), the first maximum may not be easily computed in general. However, using the decomposition of $\boldsymbol{A}_f(\boldsymbol{p})$ proposed in Eq. (2.10) it is written

$$\begin{aligned}
\max_{\substack{\boldsymbol{p}\in\mathbb{P}\\ \hat{\boldsymbol{x}}\in\mathcal{X}}} (\boldsymbol{m}_i^T \boldsymbol{A}_{f,0}^{-1}\boldsymbol{A}_f(\boldsymbol{p})\hat{\boldsymbol{x}}) &= \max_{\substack{\boldsymbol{p}\in\mathbb{P}\\ \hat{\boldsymbol{x}}\in\mathcal{X}}} (\boldsymbol{m}_i^T \boldsymbol{A}_{f,0}^{-1}(\boldsymbol{A}_{f,0} + \sum_{j=1}^{N_p} p_j \boldsymbol{A}_{f,j})\,\hat{\boldsymbol{x}}) \\
&= \max_{\substack{\boldsymbol{p}\in\mathbb{P}\\ \hat{\boldsymbol{x}}\in\mathcal{X}}} (\boldsymbol{m}_i^T \boldsymbol{A}_{f,0}^{-1}\boldsymbol{A}_{f,0}\hat{\boldsymbol{x}} + \boldsymbol{m}_i^T \boldsymbol{A}_{f,0}^{-1}\sum_{j=1}^{N_p}(p_j \boldsymbol{A}_{f,j})\hat{\boldsymbol{x}}) \\
&\leq \max_{\hat{\boldsymbol{x}}\in\mathcal{X}} (\boldsymbol{m}_i^T \boldsymbol{A}_{f,0}^{-1}\boldsymbol{A}_{f,0}\hat{\boldsymbol{x}}) + \max_{\substack{\boldsymbol{p}\in\mathbb{P}\\ \hat{\boldsymbol{x}}\in\mathcal{X}}} (\boldsymbol{m}_i^T \boldsymbol{A}_{f,0}^{-1}\sum_{j=1}^{N_p}(p_j \boldsymbol{A}_{f,j})\hat{\boldsymbol{x}}) \\
&\leq \max_{\hat{\boldsymbol{x}}\in\mathcal{X}}(\boldsymbol{m}_i^T \hat{\boldsymbol{x}}) + \max_{\substack{\boldsymbol{p}\in\mathbb{P}\\ \hat{\boldsymbol{x}}\in\mathcal{X}}} (\sum_{j=1}^{N_p} \boldsymbol{m}_i^T \boldsymbol{A}_{f,0}^{-1} p_j \boldsymbol{A}_{f,j}\hat{\boldsymbol{x}})
\end{aligned}$$

since p_j is a scalar and by over-estimating the maximum of the sum by the sum of the maxima one obtains

$$\max_{\substack{\boldsymbol{p}\in\mathbb{P}\\ \hat{\boldsymbol{x}}\in\mathcal{X}}} (\boldsymbol{m}_i^T \boldsymbol{A}_{f,0}^{-1}\boldsymbol{A}_f(\boldsymbol{p})\hat{\boldsymbol{x}}) \leq \overbrace{\max_{\hat{\boldsymbol{x}}\in\mathcal{X}}(\boldsymbol{m}_i^T \hat{\boldsymbol{x}})}^{=q_i} + \sum_{j=1}^{N_p} \max_{\substack{p_j^{min}\leq p_j \leq p_j^{max}\\ \hat{\boldsymbol{x}}\in\mathcal{X}}} (p_j \boldsymbol{m}_i^T \boldsymbol{A}_{f,0}^{-1}\boldsymbol{A}_{f,j}\hat{\boldsymbol{x}}) \, . \tag{3.30}$$

By definition of the constraints describing $\mathcal{X}$ the first maximum is easily identified as being equal to q_i. The maxima computed within the sum are scaled linear programs. It is proposed to decompose these optimisation problems using the rule

$$\begin{aligned}
\max_{\substack{a_{min}\leq a \leq a_{max}\\ \boldsymbol{x}\in\mathcal{X}}} (a\,\boldsymbol{b}^T\boldsymbol{x}) = \max \{ & a_{min} \min_{\boldsymbol{x}\in\mathcal{X}}(\boldsymbol{b}^T\boldsymbol{x}),\ a_{min} \max_{\boldsymbol{x}\in\mathcal{X}}(\boldsymbol{b}^T\boldsymbol{x}), \\
& a_{max} \min_{\boldsymbol{x}\in\mathcal{X}}(\boldsymbol{b}^T\boldsymbol{x}),\ a_{max} \max_{\boldsymbol{x}\in\mathcal{X}}(\boldsymbol{b}^T\boldsymbol{x}) \}
\end{aligned}$$

which holds true because of the independent variation of a and $\boldsymbol{x}$ in seeking the optimum. Therefore each optimisation within the sum reduces to two simpler linear programs. Consequently, by

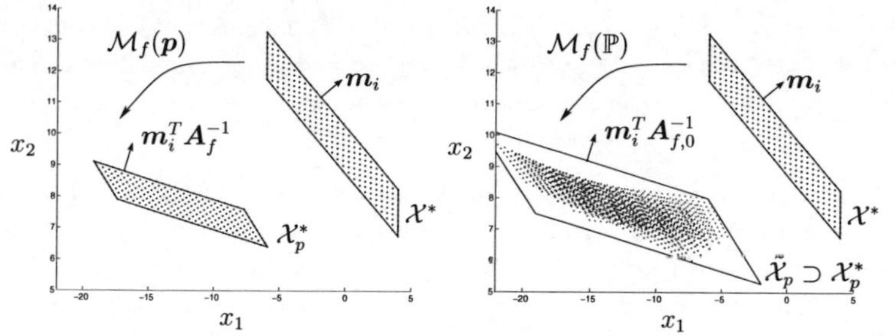

(a) Exact model case ($p = 0$). The predicted polyhedron is minimal.

(b) Uncertain model ($p \in \mathbb{P}$). The predicted polyhedron is complete.

Figure 3.9.: Predicted set for exact and uncertain models.

merging Eqs. (3.29) and (3.30) together, and using the above decomposition of the scaled linear program, it is obtained:

$$\boldsymbol{m}_i^T \boldsymbol{A}_{f,0}^{-1} \boldsymbol{x} \leq q_i + \sum_{j=1}^{N_p} \max_{\substack{p_j^{min} \leq p_j \leq p_j^{max} \\ \hat{\boldsymbol{x}} \in \mathcal{X}}} (p_j \boldsymbol{m}_i^T \boldsymbol{A}_{f,0}^{-1} \boldsymbol{A}_{f,j} \hat{\boldsymbol{x}}) + \max_{\boldsymbol{u} \in \mathcal{U}} (\boldsymbol{m}_i^T \boldsymbol{A}_{f,0}^{-1} \boldsymbol{B}_f \boldsymbol{u})$$

$$\boldsymbol{m}_i^T \boldsymbol{A}_{f,0}^{-1} \boldsymbol{x} \leq h_i \ ,$$

with h_i as defined in Eq. (3.28). Hence, $\boldsymbol{x} \in \mathcal{X}_{f,p} = \{\boldsymbol{x} \mid \boldsymbol{M}\boldsymbol{A}_{f,0}^{-1} \leq \boldsymbol{h}\}$ such that $\mathcal{X}_{f,p}^* \subset \mathcal{X}_{f,p}$. □

The fact that the exact set $\mathcal{X}_{f,p}^*$ cannot be exactly computed for uncertain models $\mathcal{M}_f(\mathbb{P})$ (using convex sets) is shown using the following numerical example. Consider an autonomous system which system matrix is known up to a single uncertain parameter $p \in \mathbb{P} = [-0.1, 0.1]$:

$$\boldsymbol{A}_f(p) = \begin{bmatrix} 0.55 & -1.20 + 2p \\ 0.25 + p & 0.80 \end{bmatrix}$$

$$\boldsymbol{A}_f(p) = \boldsymbol{A}_{f,0} + p\boldsymbol{A}_{f,1} = \begin{bmatrix} 0.55 & -1.20 \\ 0.25 & 0.80 \end{bmatrix} + p \begin{bmatrix} 0 & 2 \\ 1 & 0 \end{bmatrix} .$$

Assume that the minimal set $\mathcal{X}^*$ is known at time k and is represented as a polyhedron: $\mathcal{X}^*(k) = \{\boldsymbol{x} \mid \boldsymbol{M}\boldsymbol{x} \leq \boldsymbol{q}\}$. In Fig. 3.9, the minimal predicted set $\mathcal{X}_p^*(k+1)$ is visually approximated by considering the (real-valued) prediction of the fine grid of points covering $\mathcal{X}^*(k)$. For comparison, the polyhedra obtained in the case of a known model (on the left) and in the case of the uncertain model (on the right) are shown.

In the case of an exact model (*i.e.* $p = 0$), the predicted set exactly fits the cloud of points: the prediction from Prop. 3.2 (or Prop. 3.3) is minimal. For the uncertain model (*i.e.* $\mathbb{P} = [-0.1,\ 0.1]$), the grid of points (obtained for 5 different values of the parameter p) does not form a convex set and cannot be represented using a polyhedron. The computation proposed in Prop. 3.4 yields a larger set including all points: the result is complete.

Comparing Prop. 3.4 to the expression of the predicted set found in Alg. II, it is seen that a set of constraints $\boldsymbol{M}_p = \boldsymbol{M}\boldsymbol{A}_{f,0}^{-1}$ has been used. As proven in Theorem 4 any set of constraints could have been used, however, the idea pursued in the proposition is to preserve the information acquired over time (contained in $\boldsymbol{M}$). Indeed, with this choice of constraint orientation, and by comparing the two plots shown in Fig. 3.9, it is understood that if $\mathbb{P} \to \{\boldsymbol{0}\}$ then $\mathcal{X}_{f,p} \to \mathcal{X}^*_{f,p}$.

Remark 3.6 – The sum found in Eq. (3.28) is simpler to determine if the parameters are assumed to remain positive. In this case the scaled linear program becomes

$$\max_{\substack{0 \le a_{min} \le a \le a_{max} \\ \boldsymbol{x} \in \mathcal{X}}} (a\, \boldsymbol{b}^T \boldsymbol{x}) = \begin{cases} a_{max} \max_{\boldsymbol{x} \in \mathcal{X}}(\boldsymbol{b}^T \boldsymbol{x}), & \text{if } \max_{\boldsymbol{x} \in \mathcal{X}}(\boldsymbol{b}^T \boldsymbol{x}) > 0, \\ a_{min} \max_{\boldsymbol{x} \in \mathcal{X}}(\boldsymbol{b}^T \boldsymbol{x}), & \text{otherwise .} \end{cases}$$

While the uncertain model $\mathcal{M}_f(\mathbb{P})$ from Eqs. (2.7)–(2.9) can always be brought into a form satisfying $\mathbb{P} \subset \mathbb{R}_+^{N_p}$, this transformation modifies the value of the "main" matrix $\boldsymbol{A}_{f,0}$ in the decomposition (2.10). Therefrom the result of the predicted set (Prop. 3.4) is also modified. Once again, a tradeoff needs to be found with respect to the computation complexity.

Computing the measured set

Proposition 3.5 (Measured set) *Given the interval set of inputs $\mathcal{U}$ from Eq. (3.23), the interval set of outputs $\mathcal{Y}$ from Eq. (3.24) and the output Eq. (2.8) of a linear state-space model[4] $\mathcal{M}_f$. The measured set $\mathcal{X}^*_{f,m}$ is obtained as*

$$\mathcal{X}^*_{f,m} = F^*_{f,m}(\mathcal{Y},\mathcal{U}) = \{\boldsymbol{x} \in \mathbb{R}^n \mid \boldsymbol{N}\boldsymbol{x} \le \boldsymbol{h}\}$$

with $\boldsymbol{N} \in \mathbb{R}^{2r \times n}$ defined as

$$\boldsymbol{N} := \begin{bmatrix} \boldsymbol{C}_f \\ -\boldsymbol{C}_f \end{bmatrix}$$

[4] In this thesis, both the known models $\mathcal{M}_f(\boldsymbol{p})$ and the uncertain models $\mathcal{M}_f(\mathbb{P})$ have the same (known) output equation. Uncertainties in the $\boldsymbol{C}_f$ matrix lead to hyperboloid measured sets $\mathcal{X}_{f,m}$, [91].

and the position $\boldsymbol{h} = (h_1, \ldots, h_{2r})^T \in \mathbb{R}^{2r}$ *defined for* $i = 1, \ldots, r$ *with*

$$\begin{aligned} h_i &:= y_{i,\max} - \min_{\boldsymbol{u} \in \mathcal{U}} (\boldsymbol{d}_{f,i}^T \boldsymbol{u}) \ , \\ h_{r+i} &:= -\big(y_{i,\min} - \max_{\boldsymbol{u} \in \mathcal{U}} (\boldsymbol{d}_{f,i}^T \boldsymbol{u})\big) \ . \end{aligned}$$

For a model with no feedthrough this simplifies to $\mathcal{X}_{f,m}^* = \{\boldsymbol{x} \mid \left[\begin{smallmatrix} \boldsymbol{C}_f \\ -\boldsymbol{C}_f \end{smallmatrix}\right] \boldsymbol{x} \leq \left(\begin{smallmatrix} \boldsymbol{y}_{\max} \\ -\boldsymbol{y}_{\min} \end{smallmatrix}\right)\}$. ■

Proof: The set is constructed by considering the inequalities for each component y_i of the output measurement vector $\boldsymbol{y} \in \mathbb{R}^r$. From the model description in Eq. (2.8) it is known that $\boldsymbol{y} = \boldsymbol{C}_f \boldsymbol{x} + \boldsymbol{D}_f \boldsymbol{u}$ for an input $\boldsymbol{u} \in \mathcal{U}$. Since $\boldsymbol{y} \in \mathcal{Y}$, and $\mathcal{Y}$ is an interval set:

$$\boldsymbol{y}_{\min} \leq \boldsymbol{C}_f \boldsymbol{x} + \boldsymbol{D}_f \boldsymbol{u} \leq \boldsymbol{y}_{\max} \ .$$

The above vectorial inequalities may be rewritten on a scalar basis with $\boldsymbol{c}_{f,i}^T$ and $\boldsymbol{d}_{f,i}^T$ the i-th rows of $\boldsymbol{C}_f$ and $\boldsymbol{D}_f$:

$$y_{i,\min} \leq \boldsymbol{c}_{f,i}^T \boldsymbol{x} + \boldsymbol{d}_{f,i}^T \boldsymbol{u} \leq y_{i,\max} \ , \quad i = 1, \ldots, r \ . \tag{3.31}$$

Furthermore, since

$$\min_{\boldsymbol{u} \in \mathcal{U}} (\boldsymbol{d}_{f,i}^T \boldsymbol{u}) \leq \boldsymbol{d}_{f,i}^T \boldsymbol{u} \leq \max_{\boldsymbol{u} \in \mathcal{U}} (\boldsymbol{d}_{f,i}^T \boldsymbol{u}) \quad \text{i.e.} \quad -\max_{\boldsymbol{u} \in \mathcal{U}} (\boldsymbol{d}_{f,i}^T \boldsymbol{u}) \leq -\boldsymbol{d}_{f,i}^T \boldsymbol{u} \leq -\min_{\boldsymbol{u} \in \mathcal{U}} (\boldsymbol{d}_{f,i}^T \boldsymbol{u}) \ ,$$

the two inequalities in (3.31) are split into

$$\begin{aligned} &\boldsymbol{c}_{f,i}^T \boldsymbol{x} \leq y_{i,\max} - \boldsymbol{d}_{f,i}^T \boldsymbol{u} \leq y_{i,\max} - \min_{\boldsymbol{u} \in \mathcal{U}} (\boldsymbol{d}_{f,i}^T \boldsymbol{u}) \\ \text{and } &\boldsymbol{c}_{f,i}^T \boldsymbol{x} \geq y_{i,\min} - \boldsymbol{d}_{f,i}^T \boldsymbol{u} \geq y_{i,\min} - \max_{\boldsymbol{u} \in \mathcal{U}} (\boldsymbol{d}_{f,i}^T \boldsymbol{u}) \end{aligned}$$

from which the set description given for $\mathcal{X}_{f,m}^*$ in the proposition follows. The result for the model with no direct feedthrough is obtained by letting $\boldsymbol{D}_f = \boldsymbol{0}$ in Eq. (3.31). □

Computing the corrected set

Proposition 3.6 (Intersection of two polyhedra) *The intersection of two polyhedra* $\mathcal{X}_1$ *and* $\mathcal{X}_2$ *defined by*

$$\begin{aligned} \mathcal{X}_1 &= \{\boldsymbol{x} \mid \boldsymbol{M}_1 \boldsymbol{x} \leq \boldsymbol{q}_1\} \\ \mathcal{X}_2 &= \{\boldsymbol{x} \mid \boldsymbol{M}_2 \boldsymbol{x} \leq \boldsymbol{q}_2\} \end{aligned}$$

is the polyhedron computed as

$$\mathcal{X} = \mathcal{X}_1 \cap \mathcal{X}_2 = \left\{\boldsymbol{x} \mid \begin{bmatrix} \boldsymbol{M}_1 \\ \boldsymbol{M}_2 \end{bmatrix} \boldsymbol{x} \leq \begin{pmatrix} \boldsymbol{q}_1 \\ \boldsymbol{q}_2 \end{pmatrix} \right\} \ .$$

■

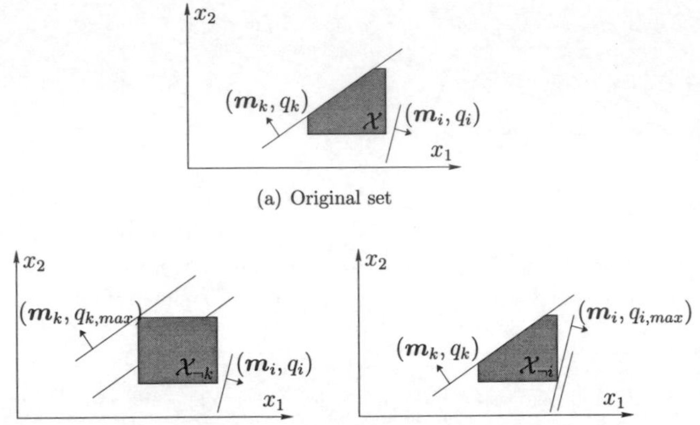

(a) Original set

(b) Non-redundant constraint: $q_{k,max} > q_k$.

(c) Redundant constraint: $q_{i,max} \le q_i$.

Figure 3.10.: Removal of redundant constraints

The above expression of the intersection can result in redundant constraints. These are searched and removed using the result from the following Prop. 3.7, illustrated in Fig. 3.10.

Proposition 3.7 (Detection of redundant constraints) *The polyhedron* $\mathcal{X} = \{\boldsymbol{x} \mid \boldsymbol{M}\boldsymbol{x} \le \boldsymbol{q}\}$ *with* s *constraints has redundant constraints if* $\exists i \in \{1, \ldots, s\}$ *such that*

$$q_{i,max} := \max_{\boldsymbol{x} \in \mathcal{X}_{\neg i}} (\boldsymbol{m}_i^T \boldsymbol{x}) \le q_i$$
$$\textit{with } \mathcal{X}_{\neg i} := \{\boldsymbol{x} \mid \boldsymbol{m}_j^T \boldsymbol{x} \le q_j,\ 1 \le j \le s,\ j \ne i\}\ .$$

If the i-th constraint is redundant it is removed from the set description without modifying the set, hence $\mathcal{X}_{\neg i} \equiv \mathcal{X}$. ∎

Therefore, an algorithm which removes the redundant constraints of a polyhedron $\mathcal{X}$ has to verify all constraints of $\mathcal{X}$ for the above redundancy condition [20, 27, 62]. In the worst case this results in computing s-LPs (LP: linear program) each of which is constrained by $q-1$ inequalities. (The computational worst case corresponds to a polytope which has no redundant constraint, hence no inequality is removed and all s optimisation problems are of equal size.)

Computing the overapproximated set with the Δ-operator

While the intersection of two polyhedra is exactly represented as a polyhedron, the resulting set is in general more complex to describe: it contains more inequalities. In order to avoid an exponential increase of the algorithm's computing time, a simplification of the set is undertaken.

Althought the Δ-operator was used in the description of earlier propositions (*e.g.* the simplified predictions), it needed not be explicitly computed and only indicated that an overapproximation of a minimal set was obtained. It is only in the last step of the complete set-observation (Alg. II) that this operator needs to be computed using the following proposition.

Proposition 3.8 (Overapproximation of polyhedron) *Given a polyhedron $\mathcal{X}$ defined by Eq.* (3.21), *the set $\tilde{\mathcal{X}}$*

$$\tilde{\mathcal{X}} = \Delta(\boldsymbol{W}, \mathcal{X}) = \{\boldsymbol{x} \in \mathbb{R}^n \mid \boldsymbol{W}\boldsymbol{x} \leq \boldsymbol{z}\}$$

overapproximates $\mathcal{X}$ to the constraints $\boldsymbol{W}$, such that $\tilde{\mathcal{X}} \supseteq \mathcal{X}$, with $\boldsymbol{W} = \left[\begin{smallmatrix} \boldsymbol{w}_1^T \\ \vdots \\ \boldsymbol{w}_\mu^T \end{smallmatrix}\right]$ and

$$z_i := \max_{\boldsymbol{x} \in \mathcal{X}} (\boldsymbol{w}_i^T \boldsymbol{x}), \quad i = 1, \ldots, \mu \,. \tag{3.32}$$

■

PROOF: For all $\boldsymbol{x} \in \mathcal{X}$, $\forall i \in \{1, \ldots, \mu\}$, $\boldsymbol{w}_i^T \boldsymbol{x} \leq \max_{\boldsymbol{x} \in \mathcal{X}} (\boldsymbol{w}_i^T \boldsymbol{x}) = z_i$. Hence, $\boldsymbol{x} \in \tilde{\mathcal{X}} \subseteq \mathcal{X}$. □

As studied in [65] the choice of orientations $\boldsymbol{W}$ used for the overapproximation has a strong impact on the quality (*i.e.* the precision) of the state-set reconstruction. The choice of overapproximation amounts to a tradeoff between the desired precision of the reconstruction and the affordable computational complexity. In this thesis, unless noted otherwise, an axis-parallel overapproximation is used with

$$\boldsymbol{W}_{\text{axis}} = \begin{bmatrix} \boldsymbol{I}_n \\ -\boldsymbol{I}_n \end{bmatrix} \in \mathbb{R}^{2n \times n}$$

which leads to interval sets $\mathcal{X}_f$. Even in such a case – where input, output and state sets are interval sets – the considered state-set observation distinguishes itself from an "interval observer" [93–95], since the intermediary sets (such as the predicted and measured sets) are polyhedric and not intervals. Therefore, a comparatively more precise observation results.

Using such symmetric constraints, $2n$ optimisations need to be solved to describe the approximated set. Alternatively, the minimum number of constraints needed to describe a bounded polyhedron is $C_n^{n+1} = n + 1$, [122]. Therefore the overapproximation always implies to solve at least $n+1$ linear programs. While this last choice implies the lowest computational burden, the price paid is a great loss in precision and in flexibility since the set is bounded by fewer constraints.

Remark 3.7 – The optimisation (3.32) may not have a solution if $\mathcal{X}$ is not a bounded polyhedron, leading possibly to $\exists i \in \{1,\ldots,\mu\}$ such that $z_i \to +\infty$ (and possibly to a trivial overapproximation $\tilde{\mathcal{X}} = \mathbb{R}^n$ if this is true for all i). From the set-observation perspective, this represents a complete loss of information about the state. While this does not brake completeness, it has to be prevented.

Overall computational cost

An overview of the size of the polyhedric sets obtained in the intermediary computation of the complete state-set observation and in the burden required to derive these sets is summarised in Table 3.2. If no overapproximation is computed (*i.e.* $\mathcal{X}_f := \mathcal{X}_{f,\cap}$) then the number of constraints increases by at least $2r$ in each time step.[5] While the computation of the overapproximation in each time step allows a constant computation time for each loop of the algorithm, it also leads to a loss of precision in the state-set observation. Thus – and without losing any of the algorithm's aforementioned properties – an alternative implementation of the algorithm which does not compute the overapproximated set *in each loop* is quite sensible. For example, the approximation may be computed only every M steps ($M > 1$) and/or only if $\mathcal{X}_{f,\cap}$ contains more than M constraints ($M > s + 2r$). Such a set-observation is pursued for comparison purposes in Section 5.2 (page 84). In this manner the quality of the state reconstruction is improved while keeping its computational cost under control.

Although the computation of the overapproximated set is costly (see table), it is emphasised that its existence permits to skip any step in charge of removing redundant constraints. Indeed, the set resulting from the overapproximation is, by construction, free of redundancies.

Remark 3.8 – The proposed implementations rules have considered the input set $\mathcal{U}$ to be a polyhedric set. Using interval sets, the computational burden is reduced since all LP($\mathcal{U}$) problems are *explicitly solved.* Indeed, using

$$\boldsymbol{c}^T\boldsymbol{u} = (c_1 \ \ldots \ c_m)^T \begin{pmatrix} u_1 \\ \vdots \\ u_m \end{pmatrix} = c_1 u_1 + \cdots + c_m u_m$$

$$\text{and } \mathcal{U} = \{\boldsymbol{u} \in \mathbb{R}^m \mid u_{1,\min} \le u_1 \le u_{1,\max}, \ \ldots \ , u_{m,\min} \le u_m \le u_{m,\max}\} \ ,$$

[5] An estimation of the increased number of constraints resulting from a Fourier-Motzkin elimination (Prop. 3.2) is difficult in general but can be exponential, depending on its implementation.

Intermediary set	Nb. constraints	Optim. for comput.
$\mathcal{X}_f(k)$	s	none (given)
$\mathcal{X}_{f,p}(k+1)$	s	s-LP($\mathcal{U}$) & $2N_p$-LP($\mathcal{X}_f$)
$\mathcal{X}_{f,m}(k+1)$	$2r$	$2r$-LP($\mathcal{U}$)
$\mathcal{X}_{f,\cap}(k+1)$	$s+2r$	none
$\mathcal{X}_f(k+1)$	$\mu = s$	μ-LP($\mathcal{X}_{f,\cap}$)

Table 3.2.: Computational burden of complete set-observation (Alg. II) using the simplified prediction from Prop. 3.4. (The notation q-LP($\mathcal{X}$) indicates q linear programs have to be solved over the constraints describing $\mathcal{X}$.)

the following LP($\mathcal{U}$) is solved as

$$\max_{\boldsymbol{u}\in\mathcal{U}}(\boldsymbol{c}^T\boldsymbol{u}) = \max_{\substack{u_{i,\min}\leq u_i\leq u_{i,\max}\\ 1\leq i\leq m}}(c_1u_1+\cdots+c_mu_m) = \sum_{i=1}^{m}\max_{u_{i,\min}\leq u_i\leq u_{i,\max}}(c_i\,u_i)\ ,$$

with

$$\max_{u_{i,\min}\leq u_i\leq u_{i,\max}}(c_i\,u_i) = \begin{cases} c_i\,u_{i,\max} & \text{if } c_i\geq 0\ ,\\ c_i\,u_{i,\min} & \text{if } c_i<0\ .\end{cases}$$

3.5. Set-membership state observability

While algorithms have been presented in this chapter to describe a set of states – or even the minimal set of states – no clear condition is affirmed for which this is possible or not. Indeed the algorithms can be applied to any linear state-space system with regular $\boldsymbol{A}_f$ matrices (its inverse is needed for the prediction step). Further results regarding the *set-observability* are described in this section.

As observability is seen as a *system property*, and not an algorithmic property, this section only describes properties referring to the minimal state-sets (hence applies to exact linear state-space systems with no direct feedthrough using Alg. III).

Bounded state-sets. The most commonly found definition of set-observability refers to the boundedness of the computed set of states, *cf.* [86,98,99]. For the minimal state-set as presented in this thesis the following theorem holds.

Theorem 6 (Boundedness of the Minimal State-Sets)
Given an exact linear state-space model $\mathcal{M}_f(\boldsymbol{p})$, if there exists an integer ν, such that

$$\operatorname{rank}\begin{bmatrix} \boldsymbol{C}_f \\ \boldsymbol{C}_f\boldsymbol{A}_f^{-1} \\ \vdots \\ \boldsymbol{C}_f\boldsymbol{A}_f^{-\nu} \end{bmatrix} = n\,, \quad \textit{with } 0 \leq \nu < n, \tag{3.33}$$

then the set of minimal final-states $\mathcal{X}_f^*(k \mid 0 \ldots k)$ is bounded for all $k \geq \nu$ (ν is the observability index *[79]).*

PROOF: The result is proven by constructing the minimal sets $\mathcal{X}_f^*(k) = \mathcal{X}_f^*(k \mid 0 \ldots k)$ over time using Alg. III. In the first step, the a priori set $\mathcal{X}_f(0 \mid -1) = \mathbb{R}^n$ is used and the output set is an interval set as in Eq. (3.24):

$$\mathcal{Y}(0) = \{\boldsymbol{y} \in \mathbb{R}^r \mid \boldsymbol{y}_{\min}(0) \leq \boldsymbol{y} = \boldsymbol{C}_f\boldsymbol{x} \leq \boldsymbol{y}_{\max}(0)\}\ .$$

Therefrom the minimal state-set is directly obtained since

$$\begin{aligned} \mathcal{X}_f^*(0) &= (\mathcal{X}(0 \mid -1) \cap \mathcal{X}_{f,m}^*(0)) = \mathcal{X}_{f,m}^*(0) \\ \text{hence } \mathcal{X}_f^*(0) &= \{\boldsymbol{x} \in \mathbb{R}^n \mid \boldsymbol{C}_f\boldsymbol{x} \leq \boldsymbol{y}_{\max}(0) \text{ and } \boldsymbol{C}_f\boldsymbol{x} \geq \boldsymbol{y}_{\min}(0)\} \\ \text{and } \mathcal{X}_f^*(0) &= \left\{\boldsymbol{x} \in \mathbb{R}^n \mid \left[\begin{smallmatrix} \boldsymbol{C}_f \\ -\boldsymbol{C}_f \end{smallmatrix}\right]\boldsymbol{x} \leq \left(\begin{smallmatrix} \boldsymbol{y}_{\max}(0) \\ -\boldsymbol{y}_{\min}(0) \end{smallmatrix}\right)\right\}\ . \end{aligned}$$

In the second loop, for $k = 1$, the prediction has to be computed. Considering the case of an autonomous system the simplified prediction of Prop. 3.3 is exact and yields

$$\mathcal{X}_{f,p}^*(1) = \boldsymbol{A}_f \cdot \mathcal{X}_f^*(0) = \left\{\boldsymbol{x} \in \mathbb{R}^n \mid \left[\begin{smallmatrix} \boldsymbol{C}_f\boldsymbol{A}_f^{-1} \\ -\boldsymbol{C}_f\boldsymbol{A}_f^{-1} \end{smallmatrix}\right]\boldsymbol{x} \leq \left(\begin{smallmatrix} \boldsymbol{y}_{\max}(0) \\ -\boldsymbol{y}_{\min}(0) \end{smallmatrix}\right)\right\}$$

and, similarly as in the first loop, the measured set is obtained as

$$\mathcal{X}_{f,m}^*(1) = \left\{\boldsymbol{x} \in \mathbb{R}^n \mid \left[\begin{smallmatrix} \boldsymbol{C}_f \\ -\boldsymbol{C}_f \end{smallmatrix}\right]\boldsymbol{x} \leq \left(\begin{smallmatrix} \boldsymbol{y}_{\max}(1) \\ -\boldsymbol{y}_{\min}(1) \end{smallmatrix}\right)\right\}\ .$$

By concatenating all constraints together the corrected set, at the intersection of the predicted and the measured sets is derived. This is also the next instance of the minimal state-set, hence

$$\mathcal{X}_f^*(1) = \mathcal{X}_{f,\cap}^*(1) = \left\{\boldsymbol{x} \in \mathbb{R}^n \mid \begin{bmatrix} \boldsymbol{C}_f\boldsymbol{A}_f^{-1} \\ -\boldsymbol{C}_f\boldsymbol{A}_f^{-1} \\ \boldsymbol{C}_f \\ -\boldsymbol{C}_f \end{bmatrix}\boldsymbol{x} \leq \begin{pmatrix} \boldsymbol{y}_{\max}(0) \\ -\boldsymbol{y}_{\min}(0) \\ \boldsymbol{y}_{\max}(1) \\ -\boldsymbol{y}_{\min}(1) \end{pmatrix}\right\}\ .$$

The same steps repeat in all loops of the algorithm: multiplying existing constraints by $\boldsymbol{A}_f^{-1}$ and adding new constraints oriented by $\pm\boldsymbol{C}_f$. Therefore

$$\mathcal{X}_f^*(k) = \left\{ \boldsymbol{x} \in \mathbb{R}^n \mid \begin{bmatrix} \boldsymbol{C}_f \boldsymbol{A}_f^{-k} \\ -\boldsymbol{C}_f \boldsymbol{A}_f^{-k} \\ \boldsymbol{C}_f \boldsymbol{A}_f^{-(k-1)} \\ -\boldsymbol{C}_f \boldsymbol{A}_f^{-(k-1)} \\ \vdots \\ \boldsymbol{C}_f \\ -\boldsymbol{C}_f \end{bmatrix} \boldsymbol{x} \leq \begin{pmatrix} \boldsymbol{y}_{\max}(0) \\ -\boldsymbol{y}_{\min}(0) \\ \boldsymbol{y}_{\max}(1) \\ -\boldsymbol{y}_{\min}(1) \\ \vdots \\ \boldsymbol{y}_{\max}(k) \\ -\boldsymbol{y}_{\min}(k) \end{pmatrix} \right\} .$$

By reordering the constraints describing the minimal state-set and letting

$$\boldsymbol{S}_r(k) = \begin{bmatrix} \boldsymbol{C}_f \\ \vdots \\ \boldsymbol{C}_f \boldsymbol{A}_f^{-k} \end{bmatrix}$$

the symmetric nature of the set appears by rewriting it as rewritten as

$$\mathcal{X}_f^*(k) = \left\{ \boldsymbol{x} \in \mathbb{R}^n \mid \begin{bmatrix} \boldsymbol{S}_r(k) \\ -\boldsymbol{S}_r(k) \end{bmatrix} \boldsymbol{x} \leq \begin{pmatrix} \boldsymbol{y}_{r,\max}(k) \\ -\boldsymbol{y}_{r,\min}(k) \end{pmatrix} \right\} \tag{3.34}$$

with $\boldsymbol{y}_{r,\max}(k)$ and $\boldsymbol{y}_{r,\min}(k)$ resulting from the appropriate reordering of the components $\boldsymbol{y}_{\max}(i)$ and $\boldsymbol{y}_{\min}(i)$, $0 \leq i \leq k$.

The symmetric polyhedron $\mathcal{X}_f^*(k)$ is bounded if and only if $\operatorname{rank}(\boldsymbol{S}_r(k)) = n$, [72]. Following Eq. (3.33), as ν is the first time index for which the matrix $\boldsymbol{S}_r$ has full rank, it follows that $\forall k \geq \nu$, $\mathcal{X}_f^*(k)$ is bounded.

If a known input is considered, *i.e.* with $\mathcal{U}(k) = \{\boldsymbol{u}^\circ(k)\}$, the derivation remains identical in its principle. If an unknown input is considered, the Fourier-Motzkin elimination has to be used. While this derivation is mathematically more complex, it only results in a larger set of states but does not alter the boundedness condition. □

A similar derivation is possible with respect to the minimal set of initial-states $\mathcal{X}_f^*(0 \mid 0 \dots k)$. In this case a boundedness condition is found depending on

$$\exists \nu \in \mathbb{N},\ 0 \leq \nu < n, \text{ such that } \boldsymbol{S}_o(\nu) = \begin{bmatrix} \boldsymbol{C}_f \\ \vdots \\ \boldsymbol{C}_f \boldsymbol{A}_f^{\nu} \end{bmatrix} = n \,. \tag{3.35}$$

The essence of the boundedness result shows that the well-known Kalman criterion – cited in Eq. (3.2) for the derivation of a real-valued state observer – implies the boundedness of the set-valued observer result. The two criteria from Eqs. (3.33) and (3.35) are equivalent if $\boldsymbol{A}_f$ is regular.

Convergence is not observability. While the boundedness of the minimal set of states is an important result, one may seek a more precise result and hence be interested in a property regarding the convergence of the state sets. Using a simple example and a proposition stating the evolution of a polytope's volume it is illustrated in the following that the apparent convergence of the final-state sets is not necessarily caused by a successful observation.

Consider the autonomous system

$$\boldsymbol{x}(k+1) = \begin{bmatrix} 0.55 & -1.20 \\ 0.25 & 0.80 \end{bmatrix} \boldsymbol{x}(k) = \boldsymbol{A}_f \boldsymbol{x}(k)$$
$$y(k) = \begin{bmatrix} 1 & 2 \end{bmatrix} \boldsymbol{x} = \boldsymbol{C}_f \boldsymbol{x}(k)$$

with a true initial state $\boldsymbol{x}^\circ(0) = (-1,\ 10)^T$ and an assumption on the measurement error of $e_y(k) = 1.5$ (constant error) such that, $\forall k \geq 0$,

$$\mathcal{Y}(k) = \{ y \in \mathbb{R} \mid y^\circ(k) - e_y \leq y \leq y^\circ(k) + e_y \} \ .$$

Since the system is autonomous, there is no input sets to consider. The minimal set-observation algorithm is used with an a priori set

$$\mathcal{X}(0\,|\,{-1}) = \{ \boldsymbol{x} \in \mathbb{R}^2 \mid -6 \leq x_1 \leq 4 \text{ and } 5 \leq x_2 \leq 15 \}$$

and the resulting state-sets[6] $\mathcal{X}^*(k \mid 0 \ldots k)$ for the sequence of output sets $\mathcal{Y}(0 \ldots k)$ are shown in Fig. 3.11(a) for $k = 0, \ldots, 9$.

The sequence of final-state sets appears to have a continually decreasing size with time. At first, this can be interpreted as a convergence related to the observation. However, looking at the sequence of initial-state sets for the same measurement sequence shown in Fig. 3.12 it is seen that past $k = 3$ there is no information which improves the observation! Therefore, the apparent convergence of the final-state sets – which is intuitive since any stable autonomous system converges to the origin – is not a true sign of set-observability. (For comparison purposes this scenario is the same as the one used to generate the three-dimensional plots in Fig. 3.4.)

The cause of this phenomenon is the convergence of the predicted sets alone, independently of the intersection with the measured sets. This is shown in the following proposition and illustrated in Fig. 3.11(b) which shows the result of a set-simulation in which the same a priori set $\mathcal{X}(0\,|\,{-1})$ is used and $\mathcal{X}_f^*(k) := \mathcal{X}_{f,p}^*(k)$ is computed.

[6] Properly speaking since $\mathcal{X}(0\,|\,{-1}) \neq \mathbb{R}^2$, the sets are not truly minimal, but since the minimal operators are used as described in Alg. III, the "$*$"-superscript is used.

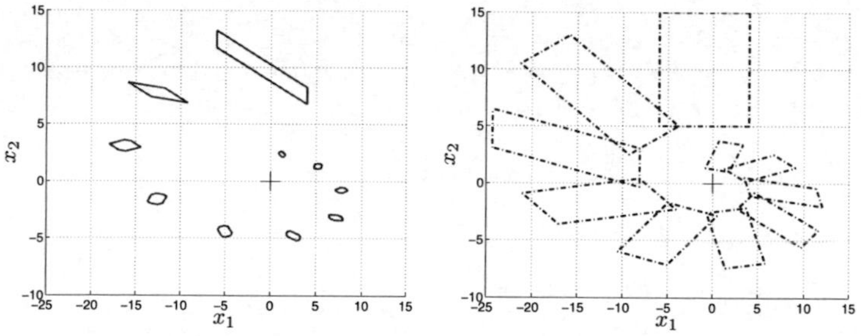

(a) Set-observation $\mathcal{X}_f^*(k \mid 0 \ldots k)$ (b) Set-simulation

Figure 3.11.: Sequence of final-state sets for $k = 0, \ldots, 9$ (autonomous system)

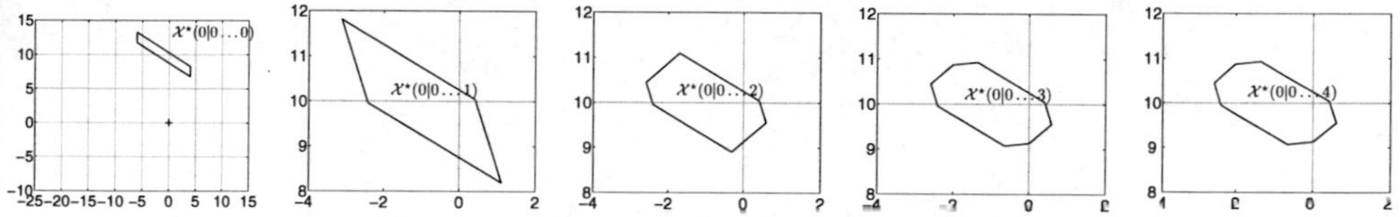

(a) $k = 0$ (b) $k = 1$ (c) $k = 2$ (d) $k = 3$ (e) $k = 4$

Figure 3.12.: Sequence of initial-state sets: for all $k \geq 3$, $\mathcal{X}^*(0\,|\,0 \ldots k) = \mathcal{X}^*(0\,|\,0 \ldots 3)$. (Scale: The axis-scaling at $k = 0$ is identical to that in Fig. 3.11. For $k \in \{1, 2, 3, 4\}$ a more detailed axis-scaling is used.)

Proposition 3.9 (Volume progression for a linear transformation) *The linear transformation of a bounded polyhedron (a polytope) $\mathcal{X}$ using the regular matrix $\boldsymbol{A}_f$ results in a polytope $\mathcal{X}'$ which volume verifies*

$$\mathrm{vol}(\mathcal{X}') = |\det(\boldsymbol{A}_f)| \; \mathrm{vol}(\mathcal{X}) \, . \qquad \blacksquare$$

PROOF: One method to compute a polytope's volume uses the triangulation of the polytope $\mathcal{X}$, *i.e.* its decomposition in simplexes, [101]. Assume that all d vertices $(\boldsymbol{p}_1, \ldots, \boldsymbol{p}_d)$ of the polytope are known and that the triangulation leads to a center $\boldsymbol{p}_0$. The volume of $\mathcal{X}$ is then expressed by the formula, [25]:

$$\mathrm{vol}(\mathcal{X}) = \frac{|\det(\boldsymbol{p}_1 - \boldsymbol{p}_0, \boldsymbol{p}_2 - \boldsymbol{p}_0, \ldots, \boldsymbol{p}_d - \boldsymbol{p}_0)|}{d!} \, .$$

As the vertices of $\mathcal{X}$ are known, the linear transformation of the polytope is computed using the $\mathcal{V}$-notation. Alternatively to Lemma 3.1 (page 48), the transformation is obtained as

$$\mathcal{X}' = \mathrm{conv}\left(\{\boldsymbol{p}_1', \; \boldsymbol{p}_2', \; \ldots\}\right)$$

with $\boldsymbol{p}'_i = \boldsymbol{A}_f \boldsymbol{p}_i,\ i = 1, \ldots, d$.

As the triangulation remains unchanged by the linear transformation, the volume of $\mathcal{X}'$ is obtained using the transformed center $\boldsymbol{p}'_0 = \boldsymbol{A}_f \boldsymbol{p}_0$ such that

$$\begin{aligned}
\mathrm{vol}(\mathcal{X}') &= \frac{|\det(\boldsymbol{p}'_1 - \boldsymbol{p}'_0, \boldsymbol{p}'_2 - \boldsymbol{p}'_0, \ldots, \boldsymbol{p}'_d - \boldsymbol{p}'_0)|}{d!} \\
&= \frac{|\det(\boldsymbol{A}_f(\boldsymbol{p}_1 - \boldsymbol{p}_0), \boldsymbol{A}_f(\boldsymbol{p}_2 - \boldsymbol{p}_0), \ldots, \boldsymbol{A}_f(\boldsymbol{p}_d - \boldsymbol{p}_0))|}{d!} \\
&= \frac{|\det(\boldsymbol{A}_f) \cdot \det(\boldsymbol{p}_1 - \boldsymbol{p}_0, \boldsymbol{p}_2 - \boldsymbol{p}_0, \ldots, \boldsymbol{p}_d - \boldsymbol{p}_0)|}{d!} \\
&= |\det(\boldsymbol{A}_f)|\ \mathrm{vol}(\mathcal{X}) \ .
\end{aligned}$$

□

As a result of this proposition, the sequence of predicted set converges for any stable autonomous system since

$$|\det(\boldsymbol{A}_f)| = \prod_{i=1}^{n} |\lambda_i(\boldsymbol{A}_f)| < 1 \ .$$

Set-observability definitions. In the previous example, the predicted sets $\mathcal{X}^*_{f,p}(k)$ for $k > 3$ are smaller than the measured sets $\mathcal{X}^*_{f,m}(k)$ such that

$$\mathcal{X}^*_{f,p}(k) \cap \mathcal{X}^*_{f,m}(k) = \mathcal{X}^*_{f,p}(k) \ , \quad \text{for } k > 3 \ ,$$

meaning that the sequence of output sets $\mathcal{Y}(4 \ldots 9)$ does not improve the state observation once $\mathcal{X}^*(3 \mid 0 \ldots 3)$ is known: the following final-state sets would have been found just as precisely with a set-simulation using $\mathcal{X}(0 \mid -1) = \mathcal{X}^*(3 \mid 0 \ldots 3)$ as a priori set. Therefore alternative definitions of set-observability are proposed which consider more than the apparent convergence of the state-sets.

Definition 3.5 (Strong Set-Observability) *The sequence of I/O sets* $(\mathcal{U}, \mathcal{Y})(0 \ldots \bar{k})$ *and the model* $\mathcal{M}_f$ *are said to be* strongly set-observable *if either of the following criteria hold for all* $k \in \{0, \ldots, \bar{k}\}$

$$\begin{aligned}
&\mathcal{X}^*_f(0 \mid 0 \ldots k) \subsetneq \mathcal{X}^*_f(0 \mid 0 \ldots k-1) \quad \textit{(criterion for sequence of initial-state sets)}, \\
&\textit{or } \left(\mathcal{X}^*_{f,p}(k) \cap \mathcal{X}^*_{f,m}(k)\right) \subsetneq \mathcal{X}^*_{f,p}(k) \quad \textit{(criterion for sequence of final-state sets)} \ .
\end{aligned}$$

Definition 3.6 (Limited Set-Observability) *The sequence of I/O sets* $(\mathcal{U}, \mathcal{Y})(0 \ldots \bar{k})$ *and the model* $\mathcal{M}_f$ *are said to be* limited set-observable to the precision $\boldsymbol{e}_x \in \mathbb{R}^n_+$ *if*

$$\forall \boldsymbol{x}_0 \in \mathcal{X}^*_f(0 \mid 0 \ldots \bar{k}), \quad |\boldsymbol{x}_0 - \boldsymbol{x}^\circ(0)| \leq \boldsymbol{e}_x \quad \textit{(initial-state set criterion)} \ .$$

These definitions are independent of one another, especially since the strong set-observability is a qualitative property whereas the limited set-observability is a quantitative property. While the former may not always hold, a precision $\boldsymbol{e}_x$ can always be found assuming Theorem 6 holds. In such case, as the state-sets are bounded for $\bar{k} \geq \nu$, there exists an interval set including the initial-state set such that

$$\mathcal{X}^*(0 \mid 0 \ldots \bar{k}) \subseteq \{\boldsymbol{x} \in \mathbb{R}^n \mid \boldsymbol{x}_{\min} \leq \boldsymbol{x} \leq \boldsymbol{x}_{\max}\}$$

and the vector

$$\tilde{\boldsymbol{e}}_x := (\boldsymbol{x}_{\max} - \boldsymbol{x}_{\min}) \geq 0$$

describes the limited precision of the algorithm's result since it verifies

$$\boldsymbol{e}_x \leq \tilde{\boldsymbol{e}}_x \, .$$

The worst case of this approximation is when the true state is at the center of the polyhedric state-set ($\boldsymbol{x}^\circ = \text{mid}(\mathcal{X}^*(0 \mid 0 \ldots k))$), whereas the proposed computation assumes it to be in a vertex of the interval set, hence $\tilde{\boldsymbol{e}}_x \leq 2\boldsymbol{e}_x$.

For the previous example, the autonomous system and the sequence $(\mathcal{U}, \mathcal{Y})(0 \ldots 3)$ are strongly set-observable since the initial-state criterion of Def. 3.5 holds true (see Fig. 3.12). As of the second definition, the widths of the smallest interval set including the initial-state set (Fig. 3.12(d)) lead to $\tilde{\boldsymbol{e}}_x = (3.2,\ 1.9)^T$ (since $\boldsymbol{x}_{\min} = (-2.6,\ 9.0)^T$ and $\boldsymbol{x}_{\max} = (0.6,\ 10.9)^T$). If a longer measurement sequence is considered, the strong set-observability property is lost but the limited set-observability holds, to the same precision.

The limited set-observability "converges" to the standard real-valued observation definition if the measurements are exactly known. Indeed, assume the measurements perfectly known, *i.e.* $\boldsymbol{e}_u(k) = \boldsymbol{0}$ and $\boldsymbol{e}_y(k) = \boldsymbol{0}$, such that the I/O sets only contain the true I/O, *i.e.* $\mathcal{U}(k) = \{\boldsymbol{u}^\circ(k)\}$ and $\mathcal{Y}(k) = \{\boldsymbol{y}^\circ(k)\}$. The set-observation then reconstructs exactly the true state. The two inequalities forming the final-state set at time k in Eq. (3.34) are transformed into inequalities describing the initial-state set as

$$\boldsymbol{y}_{o,\min}(k) \leq \boldsymbol{S}_o(k)\, \boldsymbol{x} \leq \boldsymbol{y}_{o,\max}(k), \qquad \text{with } \boldsymbol{S}_o \text{ as defined in Eq.(3.35).}$$

Since in this case $\boldsymbol{y}_{o,\min}(k) = \boldsymbol{y}_{o,\max}(k)$, these are equivalent to the vectorial equality

$$\boldsymbol{S}_o(k)\, \boldsymbol{x} = \boldsymbol{y}_{o,\min}(k) = \boldsymbol{y}_{o,\max}(k) = \begin{pmatrix} \boldsymbol{y}^\circ(0) \\ \vdots \\ \boldsymbol{y}^\circ(k) \end{pmatrix} .$$

Hence, the vector $\boldsymbol{x} = \boldsymbol{x}^\circ(0)$ is exactly computed if $\mathrm{rank}(\boldsymbol{S}_o(k)) = n$ (which is true for $k \geq \nu$). It results $\mathcal{X}^*(0 \mid 0 \ldots k) = \{\boldsymbol{x}^\circ(0)\}$ and therefore $\boldsymbol{e}_x = \mathbf{0}$.

This section only aims at giving guidelines in the definition of set-observability. While the presented example is autonomous, the above definitions are expressed for the general case. Noticeably, the strong set-observability is more likely to be verified for a scenario with an uncertain input ($\mathcal{U} \neq \{\boldsymbol{u}\}$) or an uncertain model since the prediction alone may not converge anymore (see Section 3.4) and yields a larger set. Other authors have studied set-observability definitions, *cf.* [32, 90, 105, 117].

4. Guaranteed diagnosis using set-observers

A diagnostic algorithm is presented based on the state-set observation. At first the relationship which ties the solution of the state-set observation (Chapter 3) together with the consistency principles (Chapter 2) is shown. Therefrom, a diagnostic algorithm is derived which achieves a complete and guaranteed diagnostic result.

4.1. Consistency-check by means of set-observation

In Chapter 2 the relationship between a model $\mathcal{M}_f$ and a set of I/O measurements $(\mathcal{U}, \mathcal{Y})(0 \ldots k)$ is given using consistency concepts (Def. 2.7). Considering distinct models corresponding to the process subject to the faults $f \in \mathbb{F}$, a set of fault candidates is obtained (Def. 2.8). This section justifies the use of state-set observation to test the model consistency and which leads to the description of a set of faults induced from the observation.

Theorem 7 (Model Consistency and Minimal Set-Observation)
The sequence of I/O sets $(\mathcal{U}, \mathcal{Y})(0 \ldots k)$ is consistent with the model $\mathcal{M}_f$ if and only if the solution to the minimal initial-state set observation $\mathcal{X}_f^*(0 \mid 0 \ldots k)$ is non-empty:

$$\mathcal{M}_f \models (\mathcal{U}, \mathcal{Y})(0 \ldots k) \iff \mathcal{X}_f^*(0 \mid 0 \ldots k) \neq \varnothing \,.$$

PROOF: The following equivalences prove the result

$$\begin{aligned}
\mathcal{X}_f^*(0 \mid 0 \ldots k) \neq \varnothing \iff & \exists \boldsymbol{x}_0 \in \mathcal{X}_f^*(0 \mid 0 \ldots k) \\
\overset{\text{Eq. (3.3)}}{\iff} & \exists (\boldsymbol{U}, \boldsymbol{Y})(0 \ldots k) \in (\mathcal{U}, \mathcal{Y})(0 \ldots k), \\
& \exists \boldsymbol{P} \in \mathbb{P}^{k+1},\ \exists \boldsymbol{x}_0 \in \mathbb{R}^n, \\
& \text{s.t. } \Psi_{\mathcal{Y}, f, \boldsymbol{P}}(\boldsymbol{x}_0, \boldsymbol{U}(0 \ldots k)) = \boldsymbol{Y}(0 \ldots k)
\end{aligned}$$

$$\overset{\text{Def. 2.7}}{\Longleftrightarrow} \mathcal{M}_f \models (\mathcal{U},\mathcal{Y})(0\ldots k) . \qquad \square$$

Corollary 4.1 *The result of Theorem 7 holds true using the minimal* final *state-sets:*

$$\mathcal{M}_f \models (\mathcal{U},\mathcal{Y})(0\ldots k) \iff \mathcal{X}_f^*(k \mid 0\ldots k) \neq \varnothing .$$

PROOF: The emptiness of the minimal initial and final state-sets have to be shown to be equivalent:

$$\mathcal{X}_f^*(0 \mid 0\ldots k) \neq \varnothing \iff \mathcal{X}_f^*(k \mid 0\ldots k) \neq \varnothing .$$

If $\mathcal{X}_f^*(0 \mid 0\ldots k) \neq \varnothing$ then $\boldsymbol{x}_0 \in \mathcal{X}_f^*(0 \mid 0\ldots k)$ such that $\exists(\boldsymbol{U},\boldsymbol{Y})(0\ldots k) \in (\mathcal{U},\mathcal{Y})(0\ldots k)$ and $\exists\boldsymbol{P} \in \mathbb{P}^{k+1}$ with $\boldsymbol{Y} = \Psi_{y,f,\boldsymbol{P}}(\boldsymbol{x}_0,\boldsymbol{U})$. A final state $\boldsymbol{x}_f(k)$ corresponds to this admissible input sequence, with $\boldsymbol{x}_f(k) = \psi_{x,f,\boldsymbol{P}}(\boldsymbol{x}_0,\boldsymbol{U}(0\ldots k-1))$. According to the description of the final state-set observation problem in Eq. (3.4), it results that $\boldsymbol{x}_f(k) \in \mathcal{X}_f(k \mid 0\ldots k)$, hence $\mathcal{X}_f(k \mid 0\ldots k) \neq \varnothing$. This reasoning holds both ways, such that the assertion $\mathcal{X}_f(k \mid 0\ldots k) \neq \varnothing$ also leads to $\mathcal{X}_f(0 \mid 0\ldots k) \neq \varnothing$, which proves the equivalence between the non-emptiness of both sets. $\square$

Corollary 4.2 *If a model is consistent with a sequence of I/O sets, then the solution to a complete state-set observation for this same model is non-empty:*

$$\mathcal{M}_f \models (\mathcal{U},\mathcal{Y})(0\ldots k) \implies \mathcal{X}_f(k \mid 0\ldots k) \neq \varnothing .$$

PROOF: If $\mathcal{M}_f \models (\mathcal{U},\mathcal{Y})(0\ldots k)$, from Corollary 4.1 it follows $\mathcal{X}_f^*(k) \neq \varnothing$. Since the observation is complete, the inclusion $\mathcal{X}_f^*(k) \subseteq \mathcal{X}_f(k)$ holds and $\mathcal{X}_f(k) \neq \varnothing$. $\square$

As a complete set observer is used, the relation given in Corollary 4.2 is not bidirectional. Indeed, the model $\mathcal{M}_f$ can be inconsistent with the I/O set sequence $(\mathcal{U},\mathcal{Y})$ such that $\mathcal{X}_f^* = \varnothing$ (as of Corollary 4.1) but the complete set $\mathcal{X}_f$ is non empty. This is caused either by the implemented computation rules (see Section 3.4) or by the overapproximation step used to simplify the observation algorithm (see *e.g.* Alg. II). Nevertheless, using a complete state-set observation, a set of faults is defined as follows and for which diagnostic properties are described in the upcoming section.

Definition 4.1 (Set of Faults Induced from a Set Observer)
The set of faults induced from a state-set observer describing sets $\mathcal{X}_f(k)$ is

$$\mathcal{F}(k) := \{f \in \mathbb{F} \mid \mathcal{X}_f(k) \neq \varnothing\} . \tag{4.1}$$

As for the state-set observation problem (compare to Eq. (3.5)), an a priori set of faults is defined. From consistency principles it is assumed that before any I/O measurement is known, any fault can occur, which is noted

$$\mathcal{F}(-1) := \mathbb{F} . \tag{4.2}$$

4.2. Diagnostic algorithms and properties

Theorem 8 (Monotonicity of the Diagnostic Result)
The set of faults induced from a complete state-set observer (Def. 4.1) diminishes monotonically over time such that

$$\mathcal{F}(-1) \supseteq \mathcal{F}(0) \supseteq \cdots \supseteq \mathcal{F}(k) \supseteq \mathcal{F}(k+1) .$$

PROOF: By construction of the state-sets, $\mathcal{X}_f(k \mid 0 \ldots k) \neq \varnothing$ is only possible if the previous set of states is also non-empty, *i.e.* $\mathcal{X}_f(k-1 \mid 0 \ldots k-1) \neq \varnothing$. Thus, $\forall f \in \mathcal{F}(k)$ the relation $f \in \mathcal{F}(k-1)$ also holds which proves the inclusion. (This result may also be seen as a consequence of Theorem 3 and Corollary 4.2.) □

Corollary 4.3 *The set of faults induced from a complete state-set observer (Def. 4.1) is equivalently described by the recursion*

$$\mathcal{F}(k) := \{f \in \mathcal{F}(k-1) \mid \mathcal{X}_f(k \mid 0 \ldots k) \neq \varnothing\} . \tag{4.3}$$

This set contains all faults which models result in a non-empty state-set. In other words – and using a terminology common to consistency methods – the models which are *proven to be inconsistent*[1] are *excluded* from the set of faults. Therefore a third (and still equivalent) description of the set of faults

$$\mathcal{F}(k) := \mathcal{F}(k-1) \setminus \{f \in \mathcal{F}(k-1) \mid \mathcal{X}_f(k \mid 0 \ldots k) = \varnothing\} . \tag{4.4}$$

The result from Corollary 4.3, together with the initialisation condition from Eq. (4.2), is of foremost importance as it allows a recursive computation of the set of faults $\mathcal{F}$ induced by a set-observer. This is done below in Alg. 4.1 (illustrated in Fig. 4.1).

[1]By negation of the Corollary 4.2: $\mathcal{X}_f(k \mid 0 \ldots k) = \varnothing \Longrightarrow \mathcal{M}_f \not\models (\mathcal{U}, \mathcal{Y})$.

Algorithm IV (Guaranteed Diagnostic Algorithm)
GIVEN:

- *A complete state-set observation algorithm "SSO"* (e.g. *Alg. II)*
- *The models* $\mathcal{M}_f$, $f \in \mathbb{F}$
- *The sequence of I/O sets* $(\mathcal{U}, \mathcal{Y})(0 \dots \bar{k})$

INITIALISATION: $k := 0$ *and* $\mathcal{F}(-1) := \mathbb{F}$

LOOP:

1. $\forall f \in \mathcal{F}(k-1)$, *compute* $\mathcal{X}_f(k)$ *using one loop of the set-observer for model* $\mathcal{M}_f$ *with* $\mathcal{X}_f(k-1)$ *the a priori set and* $\mathcal{U}(k-1)$, $\mathcal{U}(k)$ *and* $\mathcal{Y}(k)$ *the needed I/O sets:*

$$\mathcal{X}_f(k-1),\ \mathcal{U}(k-1),\ \mathcal{U}(k),\ \mathcal{Y}(k) \quad \xrightarrow[(Alg.\ II)]{SSO} \quad \mathcal{X}_f(k)$$

2. *Determine the induced set of faults*

$$\mathcal{F}(k) := \{f \in \mathcal{F}(k-1) \mid \mathcal{X}_f(k) \neq \varnothing\}.$$

3. *If* $k < \bar{k}$, *then* $k := k + 1$ *and go to Step 1.*

RESULT: *The sequence of induced fault sets* $\mathcal{F}(k)$, $0 \leq k \leq \bar{k}$.

Theorem 9 (Complete and Guaranteed Diagnosis)
The set of faults $\mathcal{F}$ *resulting from Alg. IV is a complete and guaranteed diagnosis,* i.e.

$$\forall k \in [0, \bar{k}], \quad \mathcal{F}(k) \supseteq \mathcal{F}^*(k) \quad \text{and} \quad f^\circ \in \mathcal{F}(k)\ .$$

PROOF: By Def. 2.8 (set of fault candidates), it is known that $\forall f \in \mathcal{F}^*(k)$, $\mathcal{M}_f \models (\mathcal{U}, \mathcal{Y})(0 \dots k)$. Following Corollary 4.2, $\mathcal{X}_f(k) \neq \varnothing$ holds and therefore $f \in \mathcal{F}(k)$ with $\mathcal{F}$ defined in Eq. (4.1). Therefore $\mathcal{F}(k) \supseteq \mathcal{F}^*(k)$. The other results of the theorem follow directly from Theorem 1 which states that $f^\circ \in \mathcal{F}^*(k)$. □

Remark 4.1 – Step 1 of Algorithm IV uses the recursiveness of the state-set observation to efficiently compute $\mathcal{X}_f(k)$ using the previous result $\mathcal{X}_f(k-1)$. This decomposition was shown in Chapter 3 and avoids reconsidering the entire time horizon $0 \dots k$ in each loop of the diagnosis.

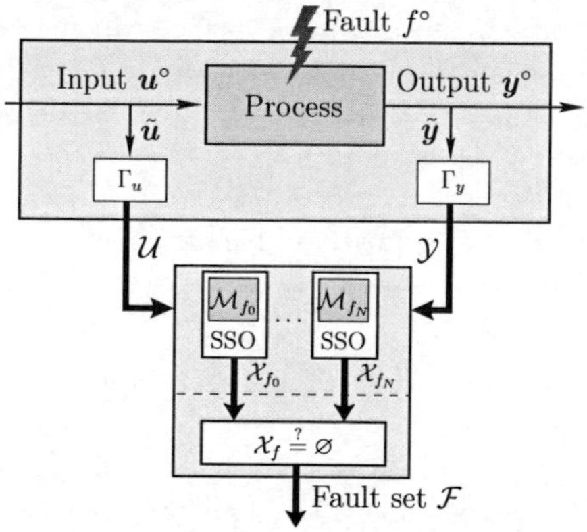

Figure 4.1.: Diagnostic setup using state-set observation (SSO).

Theorem 10 (Minimal Diagnosis)
The set of faults $\mathcal{F}$ induced from Alg. IV applied using a minimal state-set observation (Alg. III), describes a minimal diagnosis, i.e.

$$\forall k \in [0, \bar{k}], \quad \mathcal{F}(k) \equiv \mathcal{F}^*(k) \quad (\textit{and hence } f^\circ \in \mathcal{F}(k)) \ .$$

To achieve a minimal state-set observation the following conditions must hold:

- *all models are exactly known:* $\mathcal{M}_f(\boldsymbol{p}), \forall f \in \mathbb{F}$,
- *no model has direct feedthrough:* $\boldsymbol{D}_f = \boldsymbol{0}, \forall f \in \mathbb{F}$.

PROOF: By means of Theorem 5, the state-sets computed within the diagnostic algorithm are minimal: $\mathcal{X}_f(k) \equiv \mathcal{X}_f^*(k \mid 0 \ldots k)$. Therefore, the induced set of faults is

$$\mathcal{F}(k) = \{f \in \mathbb{F} \mid \mathcal{X}_f^*(k) \neq \varnothing\}$$

which is equivalent to (Corollary. 4.1)

$$\mathcal{F}(k) = \{f \in \mathbb{F} \mid \mathcal{M}_f \models (\mathcal{U}, \mathcal{Y})(0 \ldots k)\} \ .$$

This is the set of fault candidates as of Def. 2.8. □

The diagnostic algorithms presented in this chapter result in sets $\mathcal{F}(\bar{k})$ containing all faults which models are consistent with the *overall sequence* of I/O sets $(\mathcal{U}, \mathcal{Y})(0 \ldots \bar{k})$.

It is important to note that the set $\mathcal{F}(\bar{k})$ does not describe the fault models consistent with the system *at time* $\bar{k}$, but those consistent with the system over the entire horizon *from* 0 *to* $\bar{k}$. If a change of behaviour occurs within the considered time-horizon, the set of faults may become empty. Indeed, in such a case there may be no single fault f_i which model is consistent with the whole sequence $(\mathcal{U}, \mathcal{Y})(0 \ldots \bar{k})$. To avoid these issues, the fault is assumed time-invariant over the considered measurement horizon (Assumption II). Nevertheless, a possible approach to deal with such a case is proposed in [6].

The implementation of the diagnostic algorithms is for the most part straightforward given the rules presented in Section 3.4 to compute the state-set observers. Only a criteria for the detection of empty polyhedric sets is additionally necessary, which is the subject of the next proposition. It is taken from [64] and therefore given without proof (see also the "Invariant Set Toolbox" for MATLAB™ from the same author).

Proposition 4.1 (Empty polyhedron test) *A polyhedric set* $\mathcal{X} = \{\boldsymbol{x} \in \mathbb{R}^n \mid \boldsymbol{M}\boldsymbol{x} \leq \boldsymbol{q}\}$ *is empty if and only if*

$$\alpha_{\min} := \min_{\boldsymbol{M}\boldsymbol{x} \leq \boldsymbol{q} + \mathbf{1}\alpha} \alpha$$

has a strictly positive solution: $\mathcal{X} = \varnothing \iff \alpha_{\min} > 0$.
Note: this is a linear program in the variable $\binom{\alpha}{\boldsymbol{x}}$. ■

4.3. Robust diagnosability

Interpreting the diagnostic result. A classification of the diagnostic result using the set of fault candidates and according to the diagnostic terminology introduced in Section 1.1 is proposed as follows.

Proposition 4.2 (Interpretation of diagnostic result) *Based on the set* $\mathcal{F}^*$ *of fault candidates, the following diagnostic assertions are made:*

- *A fault is* ***detected*** *if and only if* $f_0 \notin \mathcal{F}^*(k)$.
- *A fault* f_i *is unambiguously* ***identified*** *if and only if* $\mathcal{F}^*(k) = \{f_i\}$. ■

Corollary 4.4 *The results of Prop. 4.2 also hold for a set of faults induced by a complete state-set observer.*

PROOF: From Theorem 9, using a complete set observer results in $\mathcal{F}(k) \supseteq \mathcal{F}^*(k)$. On the one hand, if $f_0 \notin \mathcal{F}(k)$ then $f_0 \notin \mathcal{F}^*(k)$ holds as well and the fault is detected. On the other hand, if $\mathcal{F}(k) = \{f_i\}$ and as the set of candidates contains at least one fault (Theorem 1) then $\mathcal{F}^*(k) = \{f_i\}$ and the fault f_i is identified. □

The above results explain the importance given in this thesis to seek a complete diagnosis. The corollary is only possible with a complete set of faults.

Notions of diagnosability. The completeness is essential to correctly interpret the set of faults, but it has no declarative value about the *diagnosability* of the process, [36]. For example, when a fault causes the measured value from a sensor to deviate from its true value of an amplitude less than $\boldsymbol{e}_y$ (the uncertainty assumptions on the output) then this faulty behaviour cannot be distinguished from the faultless behaviour. Due to completeness, the diagnostic result is correct (both faults are in the set of candidates) but the diagnosability[2] is not given: it cannot be asserted whether the process is faultless or faulty.

Definitions and criteria for diagnosability arc fairly widespread for discrete-event systems [36, 97] but are usually more difficult to express for continuous-time systems. Definitions have been proposed [106], but criteria to verify these either lack or are limited to simple cases. A thorough survey is found in [14] which classifies existing definitions as either *intrinsic* or *performance-based.* The former consider diagnosability as a system property independent of the diagnostic method used. The latter consider a given method together with a theoretical or numerical assessment of their expected success (*e.g.* using fault signatures, incidence matrices, signal-to-noise ratio, *etc.*). The survey concentrates however on diagnostic methods using residual generation and evaluation, and is in most cases limited to an analysis of the static behaviour.

A comparison between the diagnosability of continuous systems with that of discrete-event systems is proposed in [111]. The results are far from general as the relationship is shown based on a fault signature approach using only a model of the faultless behaviour for continuous system and where the discrete-event method considers the event chronology (*i.e.* including information about the system dynamics).

As the present method is consistency-based, the definitions found for discrete-event systems [36, 97, 102] can be ported and are expressed in the following definitions.

[2]The diagnosability refers here either the detectability or to the identifiability of a fault.

Definition 4.2 (Fault Detectability) *The fault $f_i \in \mathbb{F}$, $f_i \neq f_0$, is k-*detectable *if*

$$\exists (\boldsymbol{U}, \boldsymbol{Y}) \in (\mathbb{R}^m \times \mathbb{R}^r)^k \text{ such that } \mathcal{M}_{f_i} \models (\boldsymbol{U}, \boldsymbol{Y})(0 \ldots k-1)$$
$$\text{and } \mathcal{M}_{f_0} \not\models (\boldsymbol{U}, \boldsymbol{Y})(0 \ldots k-1) \; .$$

The fault is detectable *if there exists $k \in \mathbb{N}$ for which the fault is k-detectable.*

Definition 4.3 (Fault Identifiability) *The fault $f_i \in \mathbb{F}$ is k-*identifiable *if*

$$\exists (\boldsymbol{U}, \boldsymbol{Y}) \in (\mathbb{R}^m \times \mathbb{R}^r)^k \text{ such that } \mathcal{M}_{f_i} \models (\boldsymbol{U}, \boldsymbol{Y})(0 \ldots k-1)$$
$$\text{and } \forall f_j \in \mathbb{F}, \; f_j \neq f_i, \; \mathcal{M}_{f_j} \not\models (\boldsymbol{U}, \boldsymbol{Y})(0 \ldots k-1) \; .$$

The fault is identifiable *if there exists $k \in \mathbb{N}$ for which the fault is k-identifiable.*

These definitions represent *necessary* (but not sufficient) conditions for the detection and identification in the sense of Prop. 4.2 to occur. Indeed, they state that there exists at least one (real-valued) sequence of I/O for which the modelled behaviours distinguish themselves. Therefore, these definitions are also referred to as *weak diagnosability* since it cannot be assured that the diagnosis will succeed for the currently measured sequence of I/O, [107]. This is all the more true with sequences of I/O sets.

Following this line of thought, the notion of online and offline diagnosis is described in [71, 97]. Online diagnosis represents the result achievable while the process is "at work" and hence limited in the spectrum of I/O sequences it generates (*e.g.* a car driven on the highway). For the offline diagnosis, the process is seen as on a test-bed where arbitrarily chosen control signals (the input sequences) can be forced upon the system – while possibly using additional external computer power for the diagnosis – hence possibly triggering the fault detection (*e.g.* a car being repaired in a garage). Offline diagnosis appeals since it takes place in laboratory-like conditions which are well under control. However, it is then often difficult to reproduce the faults and hence to diagnose them offline. Having each their pros and cons, both online and offline diagnostic strategies are important.

These notions of detectability and identifiability are illustrated using the modelled behaviours $\tilde{\mathcal{B}}_f$. Defined earlier in Eq. (2.1), these are now described equivalently and in a simpler way using the consistency notation, such that

$$\tilde{\mathcal{B}}_f := \{(\boldsymbol{U}, \boldsymbol{Y})(0 \ldots k) \mid \mathcal{M}_f \models (\boldsymbol{U}, \boldsymbol{Y})(0 \ldots k), \; k \in \mathbb{N}\} \; . \tag{4.5}$$

A fault f_i is then (weakly) detectable if

$$\tilde{\mathcal{B}}_{f_i} \backslash \tilde{\mathcal{B}}_{f_0} \neq \varnothing$$

and is (weakly) identifiable if

$$\tilde{\mathcal{B}}_{f_i} \backslash \Big(\bigcup_{\substack{j=0 \\ j \neq i}}^{N} \tilde{\mathcal{B}}_{f_j} \Big) \neq \varnothing \,.$$

Fig. 4.2 shows a possible layout of a fault behaviours in the universum (space of I/O sequences). The fault f_4 is not detectable (and thus not identifiable) since its behaviour cannot be distinguished from that of the faultless system. All other faults $\{f_1, f_2, f_3\}$ are detectable as of Def. 4.2. According to Def. 4.3, faults $\{f_0, f_1, f_2\}$ are identifiable. Fault f_3 is not identifiable since its behaviour cannot be distinguished from the behaviour of f_2.

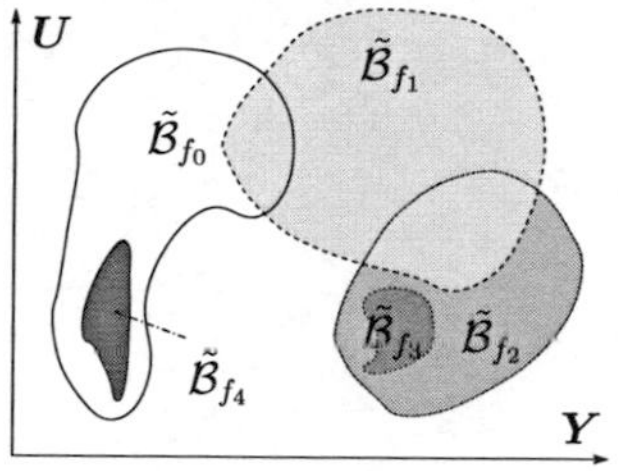

Figure 4.2.: Detectability and identifiability using modelled behaviours

In the case of observation-based diagnostic methods, a further approach to diagnosability consists in considering the observability property. For example, consider a system for which the fault is additive such that both the faultless and faulty system are modelled by the single state-space model $\mathcal{M}$

$$\begin{aligned} \boldsymbol{x}(k+1) &= \boldsymbol{A}\boldsymbol{x}(k) + \boldsymbol{B}\boldsymbol{u}(k) + \boldsymbol{F}_x f \\ \boldsymbol{y}(k) &= \boldsymbol{C}\boldsymbol{x}(k) + \boldsymbol{D}\boldsymbol{u}(k) + \boldsymbol{F}_y f \,. \end{aligned}$$

This system is diagnosed by considering an extended state-space model with the new state-variable $\tilde{\boldsymbol{x}} := \binom{\boldsymbol{x}}{f}$ such that $\mathcal{M}$ is equivalently described as (for a constant fault, hence $\dot{f} = 0$)

$$\begin{aligned} \tilde{\boldsymbol{x}}(k+1) &= \begin{bmatrix} \boldsymbol{A} & \boldsymbol{F}_x \\ 0 & 0 \end{bmatrix} \tilde{\boldsymbol{x}}(k) + \begin{bmatrix} \boldsymbol{B} \\ 0 \end{bmatrix} \boldsymbol{u}(k) = \tilde{\boldsymbol{A}}\tilde{\boldsymbol{x}} + \tilde{\boldsymbol{B}}\boldsymbol{u} \\ \boldsymbol{y}(k) &= \begin{bmatrix} \boldsymbol{C} & \boldsymbol{F}_y \end{bmatrix} \tilde{\boldsymbol{x}}(k) + \boldsymbol{D}\boldsymbol{u}(k) = \tilde{\boldsymbol{C}}\tilde{\boldsymbol{x}} + \tilde{\boldsymbol{D}}\boldsymbol{u} \,. \end{aligned}$$

Assuming the extended model remains observable, the state-set observation of this model estimates a guaranteed interval in which f° lies. If zero is not in this interval then the behaviour cannot be faultless and a fault is detected ($f^\circ > 0$). Furthermore, if a bound $\boldsymbol{e}_{\tilde{x}} = \binom{\boldsymbol{e}_x}{e_f}$ is known for the limited set-observability (Def. 3.6), then the diagnosability of any fault $f > e_f$ is assured.

The diagnosability definitions proposed in this section help understand the underlying issues in diagnosing a system. However, there is no criteria to verify these definitions given the models $\mathcal{M}_f$, $f \in \mathbb{F}$ and the I/O set sequence $(\mathcal{U}, \mathcal{Y})(0 \ldots k)$. While algorithms to check the diagnosability of discrete-event systems have been proposed, *e.g.* [36, 107], the verification of such a property appears far more complex in the continuous-valued framework.

5. Illustrative example

The theoretical results of the previous chapters in the area of set-observation and guaranteed diagnosis are illustrated using the example of a motor driving a rod in rotation.
(The method is applied to a more advanced example in Chapter 6)

Remark 5.1 – All results in this chapter are obtained by simulating mathematical models. Consequently, the information about the true fault f°, true state $\boldsymbol{x}^{\circ}$, true input $\boldsymbol{u}^{\circ}$ and true output $\boldsymbol{y}^{\circ}$ is known. These are not used to solve the tasks considered in the following and are only shown to demonstrate the guaranteed results.

5.1. Description of a motor example

The rotation of a mechanical rod is considered in which a motor brings the rod at a desired angular position using a proportional controller, Fig. 5.1. The behaviour of the motor is described by the second-order continuous-time differential equation

$$J_0\ddot{\theta}(t) = -k_v\dot{\theta}(t) - k_p\theta(t) + T_u(t) \ , \tag{5.1}$$

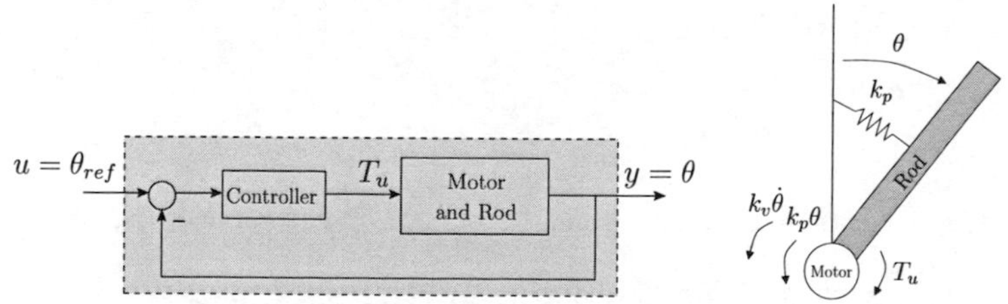

(a) Block diagram of closed-loop (b) Rotating rod

Figure 5.1.: Example of motor-driven rod

and the control law is given as

$$T_u = k\left(\theta_{ref} - \theta\right) . \tag{5.2}$$

The signals and parameters used to describe this example are found in Table 5.1. For diagnosis, different parameter values yield different dynamical models corresponding to the different fault behaviours $\mathbb{F} = \{f_0, f_1, f_2, f_3\}$ under consideration.

Signal	Description
θ	Angular position
$\dot{\theta}$	Angular velocity
$\ddot{\theta}$	Angular acceleration
T_u	Input torque
θ_{ref}	Set point for rod's angle

Parameter	Value	Description
k	10 [$\mathrm{kg\,m^2\,s^{-2}}$]	Controller gain
k_p	0.02 [$\mathrm{kg\,m^2\,s^{-2}}$]	Friction in the position
k_v	0.90 [$\mathrm{kg\,m^2\,s^{-1}}$]	Friction in the velocity (f_i, $i \neq 3$)
k_v	1.60 [$\mathrm{kg\,m^2\,s^{-1}}$]	Friction in the velocity (fault f_3)
J	0.7 [$\mathrm{kg\,m^2}$]	Inertia (faultless f_0, fault f_3)
J	1.3 [$\mathrm{kg\,m^2}$]	Inertia (fault f_1)
J	2.0 [$\mathrm{kg\,m^2}$]	Inertia (fault f_2)

Table 5.1.: Signal and parameters for the rod example

From the above equations a (continuous-time) state-space model of the closed-loop process is obtained for the state $\boldsymbol{x} = (\theta\ \dot{\theta})^T$, the input $u = \theta_{ref}$ and the output $y = \theta$:

$$\dot{\boldsymbol{x}} = \begin{bmatrix} 0 & 1 \\ \frac{-k_p-k}{J} & \frac{-k_v}{J} \end{bmatrix} \boldsymbol{x} + \begin{bmatrix} 0 \\ \frac{k}{J} \end{bmatrix} \theta_{ref} = \boldsymbol{A}_c \boldsymbol{x} + \boldsymbol{B}_c u, \tag{5.3}$$

$$y = [1\ 0]\, \boldsymbol{x} = \boldsymbol{C}_c \boldsymbol{x} = \theta, \tag{5.4}$$

$$\boldsymbol{x}(0) = \boldsymbol{x}_0. \tag{5.5}$$

The model has no feedthrough ($\boldsymbol{D}_c = \boldsymbol{0}$) and is observable for all parameter values since

$$\mathrm{rank}\left(\begin{bmatrix} \boldsymbol{C}_c \\ \boldsymbol{C}_c \boldsymbol{A}_c \end{bmatrix}\right) = \mathrm{rank}\left(\begin{bmatrix} 1 & 0 \\ 0 & 1 \end{bmatrix}\right) = 2 .$$

For the parameters of the faultless behaviour f_0, the eigenvalues are $\lambda_{1/2} = -0.64 \pm 3.73\, i$. The state sequences following a step input are shown in Fig. 5.2.

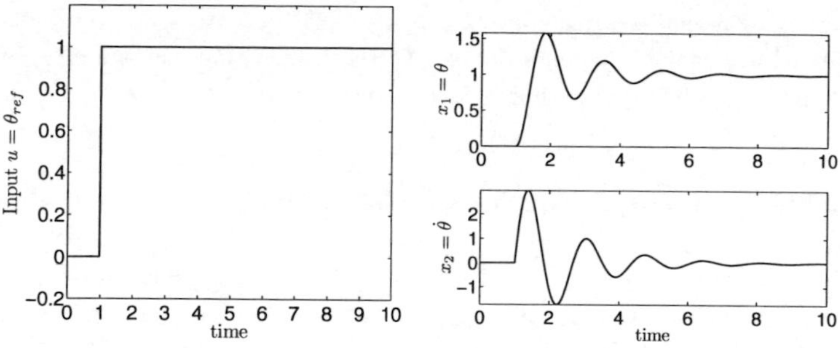

(a) Input signal u

(b) State signals $\boldsymbol{x}$

Figure 5.2.: Step response for the continuous-time model of the faultless motor ($\boldsymbol{x}_0 = \mathbf{0}$)

To apply the methods presented in this thesis, discrete-time state-space models of the fault behaviours are needed. These are obtained using zero-order-hold discretisation techniques. With a sampling time $T_s = 0.02\,[\mathrm{s}]$, the model of the faultless behaviour $\mathcal{M}_{f_0}$ is

$$\boldsymbol{x}(k+1) = \begin{bmatrix} 0.997 & 0.020 \\ -0.282 & 0.972 \end{bmatrix} \boldsymbol{x}(k) + \begin{bmatrix} 0.003 \\ 0.281 \end{bmatrix} u(k) = \boldsymbol{A}_{f_0}\boldsymbol{x}(k) + \boldsymbol{B}_{f_0}u(k), \tag{5.6}$$

$$y(k) = [1\ 0]\,\boldsymbol{x} = \boldsymbol{C}_{f_0}\boldsymbol{x}(k) = \theta, \tag{5.7}$$

$$\boldsymbol{x}(0) = \boldsymbol{x}_0. \tag{5.8}$$

The models for the faulty behaviours are obtained by discretising Eqs. (5.3)–(5.4) in which fault-specific parameter values are taken according to Table 5.1. The discrete-time models for the faulty behaviours are not shown here. The considered faults are:

- Faultless behaviour f_0.
  The behaviour is described above and its discrete-time model is $\mathcal{M}_{f_0}$.
- Fault f_1 is a small increase of the rod's inertia.
  The behaviour for the fault f_1 is considered not exactly known and is approximated by the model with uncertain parameter $J \in [0.9,\ 1.3]\,[\mathrm{kg\,m^2}]$. The corresponding discrete-time model is noted $\mathcal{M}_{f_1}(\mathbb{P})$ to emphasise the presence of uncertainties. This model includes the true fault behaviour obtained for $J^{\circ} = 1.3\,[\mathrm{kg\,m^2}]$.
- Fault f_2 is a large increase of the rod's inertia.
  The behaviour for the fault f_2 is exactly known and is obtained for $J = 2.0\,[\mathrm{kg\,m^2}]$. Its discrete-time model is $\mathcal{M}_{f_2}$.

- Fault f_3 represents a stronger rotational friction.
  The behaviour for the fault f_3 is exactly known and is obtained for $k_v = 1.6\,[\mathrm{kg\,m^2\,s^{-1}}]$. Its discrete-time model is $\mathcal{M}_{f_3}$.

5.2. State-set observation of the motor

The state-set observation of the motor is pursued using the algorithms presented in Chapter 3. In the following the model is supposed exactly known and only the faultless case $\mathcal{M}_{f_0}$ is considered.

An experiment is pursued in which the true input sequence is given as

$$u^{\circ}(k) = \begin{cases} 0 & \text{for } 0 \leq k\,T_s \leq 2.2\,[\mathrm{s}] \\ 2.2 & \text{for } 2.2\,[\mathrm{s}] < k\,T_s \leq 4.4\,[\mathrm{s}] \\ 1.0 & \text{for } k\,T_s > 4.4\,[\mathrm{s}] \end{cases} \tag{5.9}$$

and the true initial state is

$$\boldsymbol{x}^{\circ}(0) = \begin{pmatrix} 0.2 \\ 0.1 \end{pmatrix} .$$

The measured input and output sets drawn in Fig. 5.3 are obtained for the assumed uncertainty upper bounds

$$\boldsymbol{e}_u(k) = 10\%\,|u(k)| \qquad \text{(relative uncertainty)} \tag{5.10}$$

$$\text{and } \boldsymbol{e}_y(k) = 0.25\,. \qquad \text{(absolute uncertainty)} \tag{5.11}$$

In the following, the state-sets $\mathcal{X}_f(\cdot) \subset \mathbb{R}^n$ resulting from the state-set observation are polytopes of possibly complex shape. To graphically illustrate the observation's result, these state-sets are projected on each state-space dimension resulting on as many intervals as state variables (Fig. 5.4). The set-observation's result is drawn as the sequence of these intervals over time. While this represents an overapproximation of the actual state-sets, it is only used for the plots. Examples of such plots follow in Figs. 5.5(a) through 5.7(a). (In the case of an axis-parallel overapproximation, the intervals then exactly represent the state-sets.)

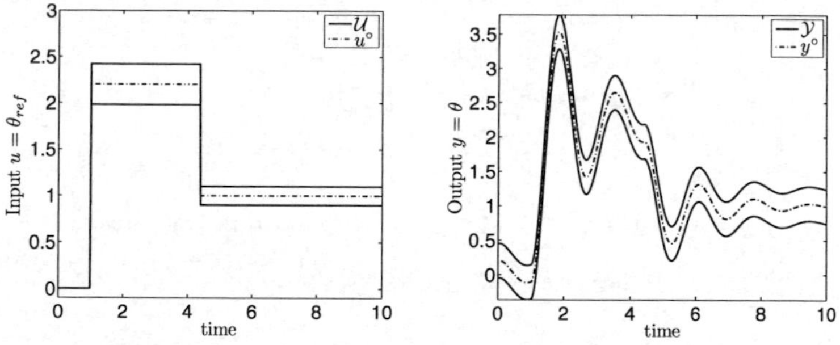

(a) Input (b) Output

Figure 5.3.: Input and output sets for exact measurement of the faultless behaviour (solid lines: edges of the interval sets, dashed lines: true measurements)

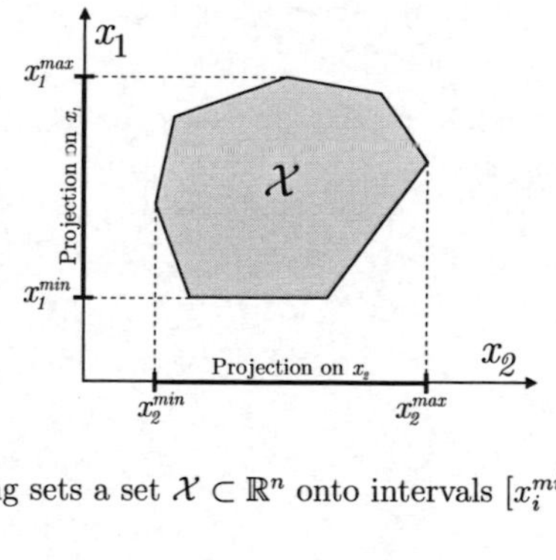

Figure 5.4.: Projecting sets a set $\mathcal{X} \subset \mathbb{R}^n$ onto intervals $[x_i^{min}, x_i^{max}]$, $i = 1, \ldots, n$.

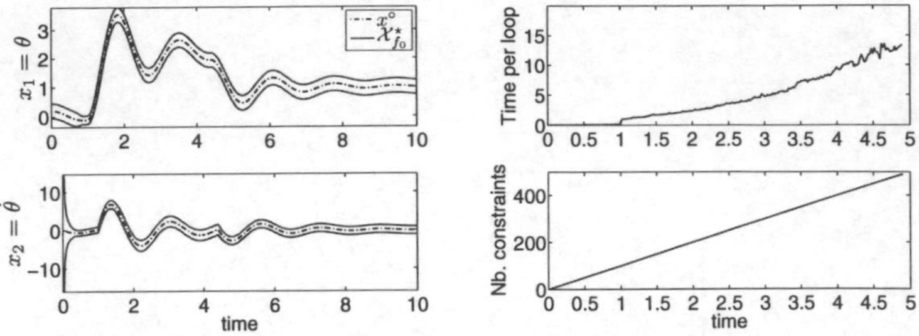

(a) State-sets $\mathcal{X}^*_{f_0}(k \mid 0 \dots k)$

(b) Computational complexity

Figure 5.5.: Minimal state-set observation

Minimal state-set observation. The model has no direct feedthrough and is exact, hence a minimal state-set observation can be pursued using Alg. III. The resulting sets $\mathcal{X}^*_{f_0}(k)$ are drawn in Fig. 5.5(a) and illustrate Theorem 2 since at all time the computed state-sets contain the true state: $\boldsymbol{x}^\circ(k) \in \mathcal{X}^*_{f_0}(k)$.

This state-set observation is computationally expensive. The 10 [s] of data (representing $10/T_s = 500$ samples) is processed in 45 minutes.[1] The high computation time is seen in the upper half of Fig. 5.5(b): the computation time required for each loop of the observation algorithm increases exponentially with time.

This effect is due to the relatively simple algorithm used here which does not remove redundant constraints. As seen in the lower half of the figure, the number of constraints needed to represent $\mathcal{X}^*_{f_0}(k)$ constantly increases, causing the solution of the linear programs to become increasingly time consuming.

While the overall computation time for this minimal state-set observation can be improved with a more advanced algorithm, its complexity is intrinsic to the problem being solved and cannot be entirely avoided.

[1] All computation times are based on an Intel P4 processor at 1.6 [Ghz] with 768 [Mb] of RAM, running MATLAB™ R14 for Linux.

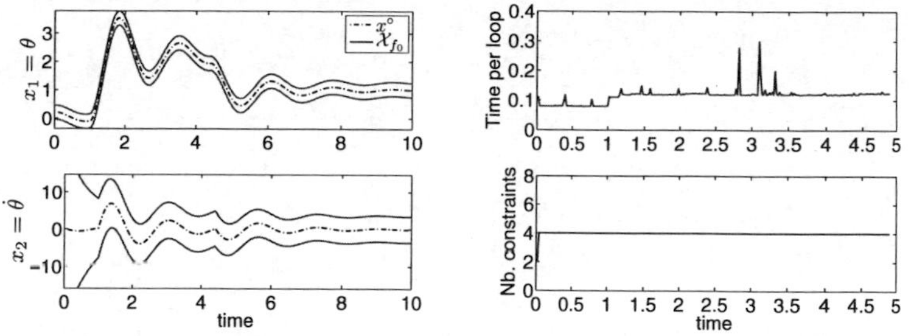

(a) State-sets $\mathcal{X}_{f_0}(k \mid 0 \ldots k)$

(b) Computational complexity

Figure 5.6.: State-set observation with axis-parallel overapproximation (in each loop)

Complete state-set observation (axis-parallel). A complete state-set observation is pursued using Alg. II. The same model $\mathcal{M}_{f_0}$ and I/O sequence are used. An axis-parallel overapproximation is considered, hence the final-step of the algorithm approximates the state-sets to the smallest interval sets. This is achieved using the Δ-operator with

$$\boldsymbol{W} = \begin{bmatrix} -\boldsymbol{I}_2 \\ \boldsymbol{I}_2 \end{bmatrix} = \begin{bmatrix} -1 & 0 \\ 0 & -1 \\ 1 & 0 \\ 0 & 1 \end{bmatrix} . \tag{5.12}$$

The resulting state-sets $\mathcal{X}_{f_0}(k)$ are drawn in Fig. 5.6(a) and illustrate Theorem 4: the sets computed by the algorithm are complete since the relation $\mathcal{X}_{f_0}(k) \supset \mathcal{X}^*_{f_0}(k)$ holds at all time (compare with the results of Fig. 5.5(a)). Consequently, the sets include the true state: $\boldsymbol{x}^\circ(k) \in \mathcal{X}_{f_0}(k)$.

This observation is evidently less precise than the minimal observation (the resulting sets are larger) but is much faster to compute. The sequence of I/O sets is treated in roughly one minute. Fig. 5.6(b) shows the time spent to compute each loop and the number of constraints describing the state-sets (beware of the different axis scaling when comparing with Fig. 5.5(b)).

Because of the overapproximation, the mathematical complexity of the sets remain constant over time: the 4 constraints' orientations from Eq. (5.12) describe the state-sets at all time k and therefore the computational complexity in each loop of the algorithm remains constant.[2]

[2]It can be seen that the very first set $\mathcal{X}_{f_0}(0)$ has only 2 constraints. It was not overapproximated using $\boldsymbol{W}$ to avoid the issue pointed out in Remark 3.7 (page 58).

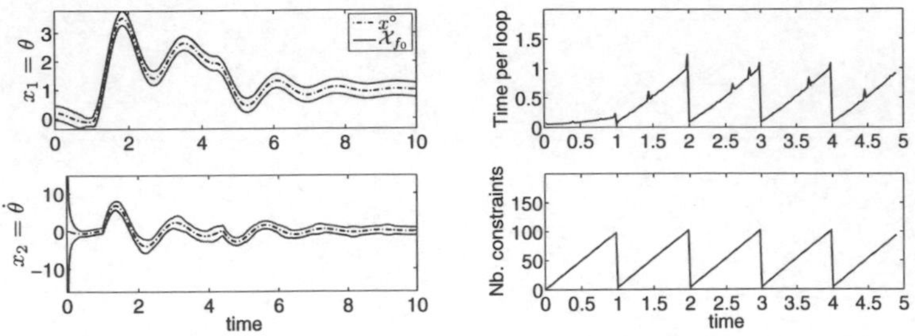

(a) State-sets $\mathcal{X}_{f_0}(k \mid 0 \ldots k)$ (b) Computational complexity

Figure 5.7.: State-set observation with axis-parallel overapproximation (every 50 loops)

Alternative state-set observation (approximation every M loops). A variation of Alg. II is proposed, in which the axis-parallel overapproximation is not computed in each loop but only every M loops, $M > 0$. The algorithm computes guaranteed and complete state-sets. The set-observation is pursued once more (with same model and I/O sets) but with this alternative algorithm using $M = 50$ (which means an overapproximation every $50 \times T_s = 1$ [s]). The result of this set-observation is shown in Fig. 5.7 and is computed in four minutes.

As evoked in Section 3.4 (see final paragraph about the computational cost of the algorithm), computing an overapproximated set reduces the problem related to redundant inequalities by "resetting" the number of constraints at regular time instants (see Fig. 5.7(b)).

This observation shows the versatility in designing alternative set-observers with interesting properties for the diagnosis. Indeed the more precise the observation is, the faster inconsistencies are discovered (further discussed in Section 5.3). In the present case the algorithm runs ten times faster than Alg. III and results in a quasi-minimal set of states (compare with Fig. 5.5).

Improved measurement using set-observation. The set-observations shown in the previous pages distinguish themselves mainly in the narrowness of the interval estimating the second state x_2. It appears surprising that the estimation of x_1 is equal for all algorithms (*i.e.* equally bad). Keeping in mind Prop. 3.9, one expects at least the minimal state-sets to converge over time, leading to a better estimation of all states, including x_1.

This effect is studied in [4] by decomposing the computation of the predicted set into the three consecutive steps:

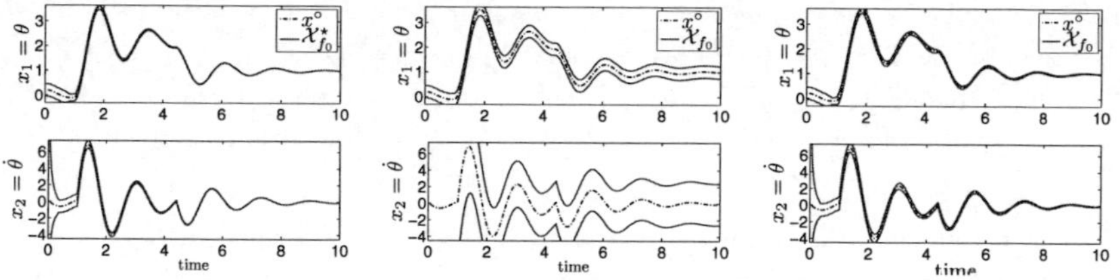

(a) Minimal observation (b) Axis-parallel approximation in every loop (c) Axis-parallel approximation only every 50 loops

Figure 5.8.: State-set observations of the example for an exact input signal

1. prediction for an autonomous movement (no input),
2. translation for the real-valued input $\bar{u}(k) = \text{mid}(\mathcal{U}(k))$,
3. translation for the uncertain input $\boldsymbol{u} \in \mathcal{U}_\Delta(k)$ with

$$\mathcal{U}_\Delta(k) := \mathcal{U}(k) - \bar{\boldsymbol{u}} := \{\boldsymbol{u} \in \mathbb{R}^m \mid \boldsymbol{u} = \hat{\boldsymbol{u}} - \bar{\boldsymbol{u}},\ \hat{\boldsymbol{u}} \in \mathcal{U}(k)\} .$$

The first step leads to a decrease in volume of the set of states (for stable systems, as of Prop. 3.9). The second step does not modify the volume of the set. Finally the third step causes an increase in volume if the input is uncertain. Depending on the amplitude of the uncertainty $\boldsymbol{e}_u$ and the model matrices, this increase can annihilate the decrease resulting from the first step. This is the case in the presented example, which explains why in all Figs. 5.5(a) through 5.7(a), the interval reconstructed for x_1 is equal to the information originally available from the measured output $\mathcal{Y}$ (compare to Fig. 5.3(b)).

To illustrate the opposite effect, the same three state-set observations are now pursued by assuming the input measurement exactly known, hence $\mathcal{U}(k) = \{u^\circ(k)\}$ and the interval sets from Fig. 5.3(a) reduce to the dashed line. All three results are shown in Fig. 5.8. These results illustrate the convergence of the minimal state-sets to the true state ($\mathcal{X}_f^* \to \{\boldsymbol{x}^\circ\}$) in the case of a stable system with no input or model uncertainties. A new scale of the x_2 axis is used to emphasise the –small but existing– difference between the minimal observation (on the left) and the complete observation (on the right).

The true output is originally only known to belong to the measured set ($y^\circ(k) \in \mathcal{Y}(k)$). Since $y^\circ(k) = x_1^\circ(k)$, the true output can also be estimated as the projection of $\mathcal{X}_{f_0}(k)$ on the x_1 dimension, hence

$$y^\circ(k) \in \{x_1 \in \mathbb{R} \mid \left(\begin{smallmatrix} x_1 \\ x_2 \end{smallmatrix}\right) \in \mathcal{X}_{f_0}(k)\}$$

which is a smaller interval. This "improved measurement" principle is seen in the first and third plots of Fig. 5.8 since the intervals containing $y^\circ = x_1^\circ$ are smaller than the original output set $\mathcal{Y}$ (compare to Fig. 5.3(b)). This effect is not seen in the second plot, since the overapproximation in every loop causes a too large loss of observation accuracy.

Effect of erroneous measurement on the set-observation. The robustness of the observation in the case of erroneous measurements is shown in the following example. (A case of erroneous model is considered in the next section with a diagnosis example.)

Up to now the state-set observations considered uncertainties on the measurements but these were not subject to any errors such that

$$\boldsymbol{u}^\circ(k) = \mathrm{mid}(\mathcal{U}(k))$$
$$\text{and } \boldsymbol{y}^\circ(k) = \mathrm{mid}(\mathcal{Y}(k))$$

held at all time instants k. This symmetry of the input set about the true input and that of the output set about the true output causes the true state to also be the center of the reconstructed state-sets, *i.e.* $\boldsymbol{x}^\circ = \mathrm{mid}(\mathcal{X}_{f_0}(k))$ as for example in Fig. 5.8(b). This is well seen in the previous results where the true state (dashed line) lies in the middle of the intervals delimiting the state-set (solid lines).

Lets consider the measurements subject to errors. To respect Assumption III, the errors are chosen smaller than the bounds from Eqs. (5.10)–(5.11). For example the input and output are erroneously measured as

$$\tilde{\boldsymbol{u}}(k) = \boldsymbol{u}^\circ(k) + \boldsymbol{d}_u(k) = \boldsymbol{u}^\circ(k) + 0.07\,|u(k)| \qquad \text{(relative error)}$$
$$\text{and } \tilde{\boldsymbol{y}}(k) = \boldsymbol{y}^\circ(k) + \boldsymbol{d}_y(k) = \boldsymbol{y}^\circ(k) - 0.15\;. \qquad \text{(absolute error)}$$

The true input and output values are not at the center of the measured sets anymore (see Fig. 5.9). As a result of which, the true state is also not at the center of the reconstructed state-sets (see Fig. 5.10). This illustrates the robustness of the set-observers with respect to unknown-but-bounded uncertainties: Despite the measurement errors, the observations are nevertheless guaranteed such that the bounds of the state-sets (solid line) include the true state (dashed line) at all times, hence $\boldsymbol{x}^\circ(k) \in \mathcal{X}_{f_0}(k)$.

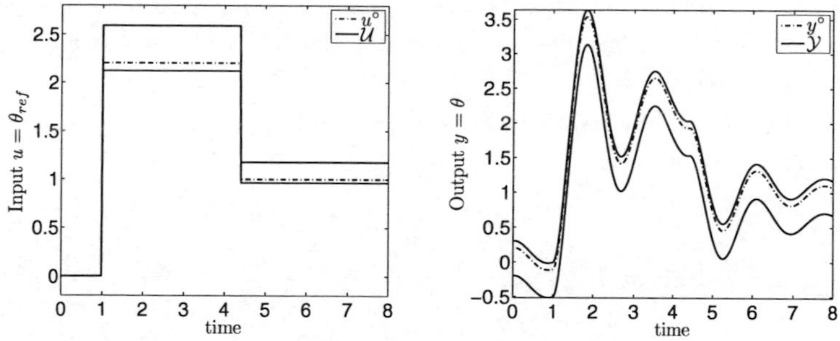

(a) Input set and true input

(b) Output set and true output

Figure 5.9.: Input and output sets for erroneous measurements of the faultless behaviour

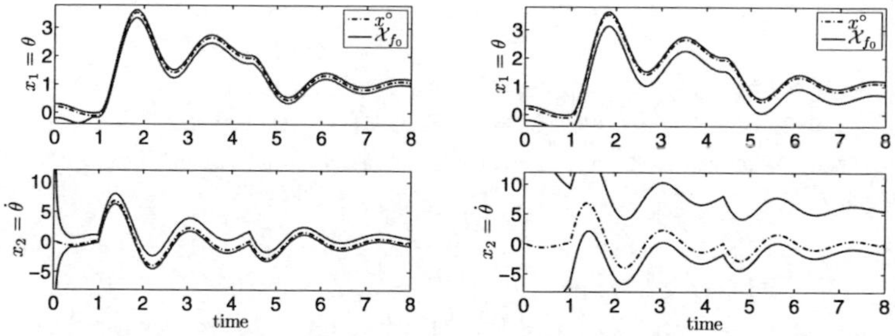

(a) Minimal set-observation $\mathcal{X}^*_{f_0}$

(b) Axis-parallel approximation $\mathcal{X}_{f_0}$

Figure 5.10.: State-set observation for erroneous I/O measurements

5.3. Diagnosis of the motor

The ultimate aim of deploying state-set observation is to check inconsistencies of models with I/O sets as described in Chapter 4 (Theorem 7). This section illustrates the proposed diagnostic method (Alg. IV) using the motor example. The faults and their models are described in Section 5.1.

All diagnostic results shown in this section are induced by a complete set-observer with axis-parallel overapproximation in each loop.

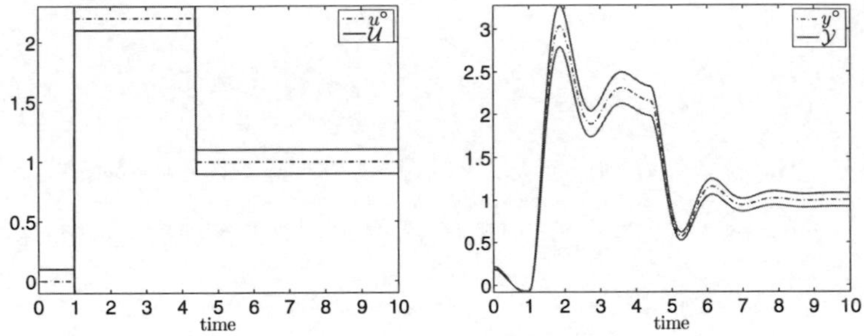

(a) Input (b) Output

Figure 5.11.: Input and output sets for the behaviour $f^\circ = f_3$ (solid lines: edges of the interval sets, dashed lines: true measurements).

Fault detection ($f^\circ = f_3$). The previous section has shown state-set observation results obtained for the model of the faultless behaviour $\mathcal{M}_{f_0}$ together with I/O corresponding to this behaviour. The state-sets were never empty proving the consistency of f_0 with the measurements.

A case of fault detection is shown. For this purpose, the model $\mathcal{M}_{f_0}$ alone is sufficient and sequences of measurements corresponding to the fault $f^\circ = f_3$ (increased friction) are used. For the same input scenario (5.9) and the following upper bounds for the measurement uncertainties

$$\boldsymbol{e}_u(k) = 0.10 \; , \tag{5.13}$$

$$\text{and } \boldsymbol{e}_y(k) = 8\% \, |y(k)| \; , \tag{5.14}$$

the sequence of I/O sets are given in Fig. 5.11. Compared to the faultless behaviour (Fig. 5.3), the process is visibly more damped due to the increased friction.

The complete state-sets observed for this sequence of I/O sets and for the model of the faultless behaviour $\mathcal{M}_{f_0}$ are shown in Fig. 5.12(a). As expected, the sets $\mathcal{X}_{f_0}$ vanish at $k_\perp = 85$ (or $t_\perp = 1.68$ [s]). Therefore $f_0 \notin \mathcal{F}(k)$, $\forall k \geq k_\perp$, and based on Prop. 4.2 a *fault is detected* because of the inconsistency of the I/O with the model of the faultless behaviour.

Fault identification ($f^\circ = f_3$). The fault identification is pursued for the same sequence of I/O sets, such that the models for all four behaviours $f \in \mathbb{F}$ are considered. As described in Alg. IV, a complete state-set observation is computed for each available fault model (The sets $\mathcal{X}_{f_0}$ were already computed for the fault detection).

The results of the bank of set-valued observers are shown in Fig. 5.12. The state-sets $\mathcal{X}_f(k)$, $f \in \mathbb{F}$, are then used to determine the consistency of the I/O with the models which results in the table:

Fault	Description	Diagnosis
f_0	Faultless behaviour	Inconsistency at $t_\perp = 1.66$ [s]
f_1	Small inertia increase	Inconsistency at $t_\perp = 1.06$ [s]
f_2	Large inertia increase	Inconsistency at $t_\perp = 0.88$ [s]
f_3	Stronger friction	No inconsistency

Therefrom, the set of faults induced from the observations is:

$$\mathcal{F}(k) = \begin{cases} \{f_0, f_1, f_2, f_3\} & \text{for } 0 \leq k\,T_s < 0.88\,[\text{s}] \\ \{f_0, f_1, f_3\} & \text{for } 0.88\,[\text{s}] \leq k\,T_s < 1.06\,[\text{s}] \\ \{f_0, f_3\} & \text{for } 1.06\,[\text{s}] \leq k\,T_s < 1.66\,[\text{s}] \\ \{f_3\} & \text{for } k\,T_s \geq 1.66\,[\text{s}] \end{cases}$$

which is graphically illustrated in Fig. 5.13. Until $k\,T_s = 1.66$ [s] it cannot be certified whether or not the process behaves abnormally since both the faultless (f_0) and a faulty (f_3) behaviour are candidates. The fault detection occurs when f_0 is excluded from the set of faults.

A single fault remains in the set and is hence the only possible fault explaining the measured I/O. Consequently *the fault f_3 is identified.* In this case the fault detection occurs simultaneously with the identification since f_0 is the last fault to be excluded from the fault set.

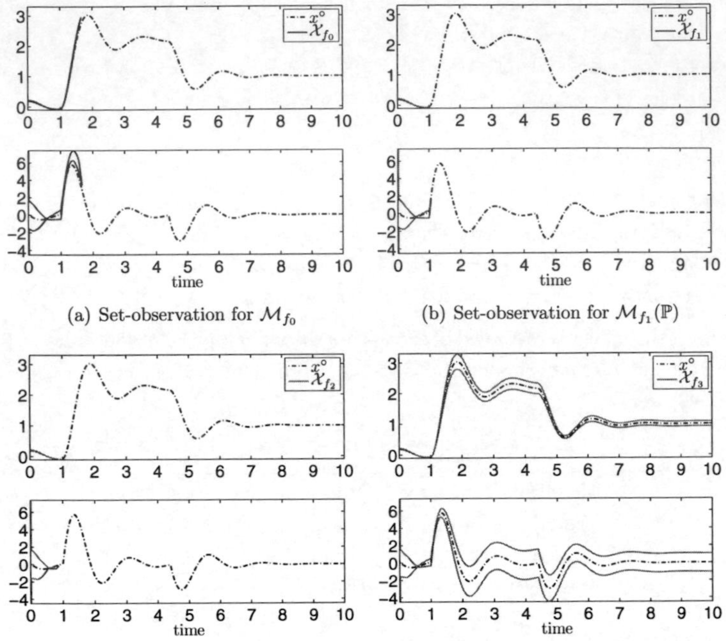

(a) Set-observation for $\mathcal{M}_{f_0}$

(b) Set-observation for $\mathcal{M}_{f_1}(\mathbb{P})$

(c) Set-observation for $\mathcal{M}_{f_2}$

(d) Set-observation for $\mathcal{M}_{f_3}$

Figure 5.12.: State-set observations for all fault models $\mathcal{M}_{f_i}$, $i \geq 0$, using the I/O sets from Fig. 5.11. All faults except f_3 are inconsistent with the measurements.

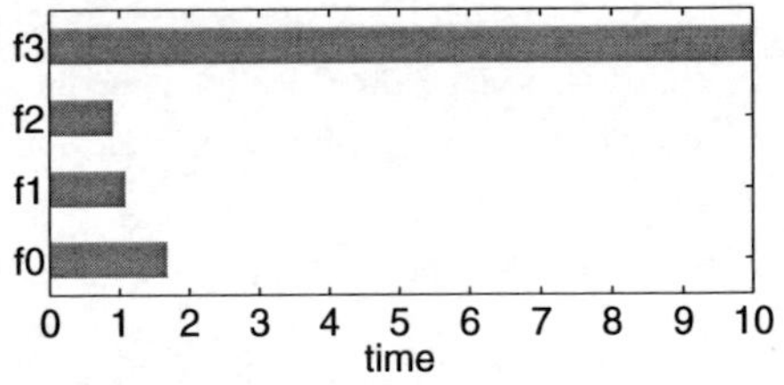

Figure 5.13.: Consistency of fault models over time, using the I/O sets from Fig. 5.11.

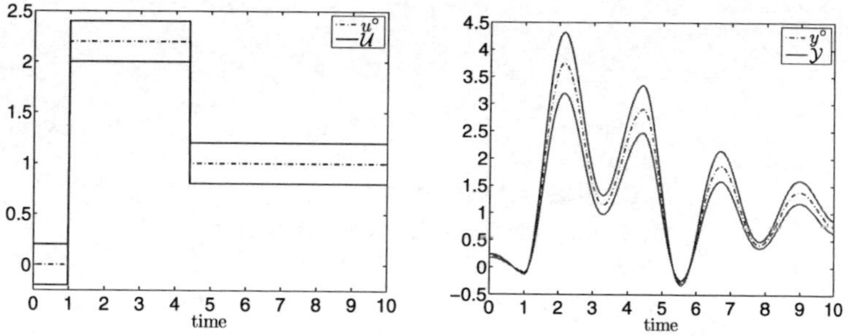

(a) Input

(b) Output

Figure 5.14.: Input and output sets for the behaviour $f^\circ = f_1$ (solid lines: edges of the interval sets, dashed lines: true measurements).

Case study with comparison exact/uncertain models ($f^\circ = f_1$). A second diagnostic experiment is undertaken to illustrate the importance of the models completeness.

Five models are considered: $\mathcal{M}_{f_0}$, $\mathcal{M}_{f_1}(\mathbb{P})$, $\mathcal{M}_{f_2}$ and $\mathcal{M}_{f_3}$ as previously and $\mathcal{M}_{f_1^{\approx}}(\boldsymbol{p})$ a fifth model obtained for the exact inertia $J = 0.9\,[\mathrm{kg\,m^2}]$. This model represents an approximation of the behaviour of fault f_1 which is not complete (and hence noted $f_1^{\approx}$).

Measurements for the fault case f_1 are used to test the diagnosis. To vary the results, the I/O sets (Fig. 5.14) are obtained for the bounds

$$\begin{aligned} \boldsymbol{e}_u(k) &= 0.20 \\ \text{and } \boldsymbol{e}_y(k) &= 15\%\,|y(k)|\ . \end{aligned}$$

The diagnostic algorithm results in the set of faults: (Fig. 5.15)

$$\mathcal{F}(k) = \begin{cases} \{f_0, f_1, f_1^{\approx}, f_2, f_3\} & \text{for } 0 \le k\,T_s < 1.12\,[\mathrm{s}] \\ \{f_1, f_1^{\approx}, f_2\} & \text{for } 1.12\,[\mathrm{s}] \le k\,T_s < 1.14\,[\mathrm{s}] \\ \{f_1, f_1^{\approx}\} & \text{for } 1.14\,[\mathrm{s}] \le k\,T_s < 5.76\,[\mathrm{s}] \\ \{f_1\} & \text{for } k\,T_s \ge 5.76\,[\mathrm{s}]\ . \end{cases}$$

A fault is detected at time $k\,T_s = 1.12\,[\mathrm{s}]$ when f_0 is excluded from the set of faults. The fault f_1 is identified at $k\,T_s = 1.14\,[\mathrm{s}]$ when all behaviours not related to f_1 are excluded. The correct diagnostic result is only obtained for the complete model: $\mathcal{M}_{f_1}(\mathbb{P})$

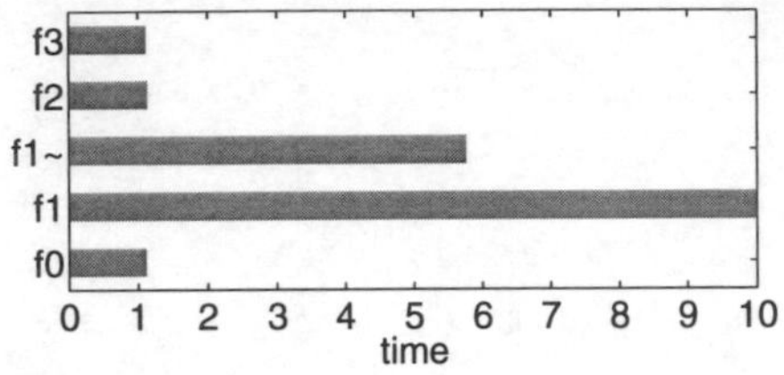

Figure 5.15.: Consistency of fault models over time using I/O sets from Fig. 5.14. Only the complete model of the behaviour of fault f_1 remains consistent ("f1∼" stands for "$f_1^{\approx}$").

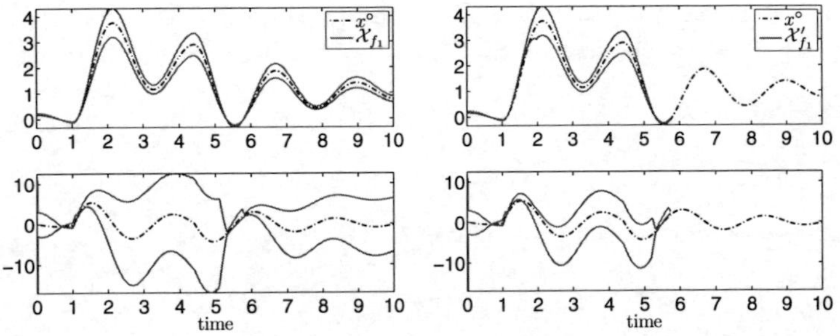

(a) State-sets $\mathcal{X}_{f_1}$ for the complete model $\mathcal{M}_{f_1}(\mathbb{P})$ with parameter uncertainties

(b) State-sets $\mathcal{X}_{f_1^{\approx}}$ for the non-complete model $\mathcal{M}_{f_1^{\approx}}(\boldsymbol{p})$ with no parameter uncertainties

Figure 5.16.: Set-observations for f_1 and $f_1^{\approx}$ resulting in the consistency of Fig 5.15.

is consistent with the whole I/O sequence whereas $\mathcal{M}_{f_1^{\approx}}(\boldsymbol{p})$ becomes inconsistent at $k\,T_s = 5.76\,[\mathrm{s}]$. The detail of the state-set observation for these two models is shown in Fig. 5.16.

This example illustrates that when considering uncertainties, the fault behaviours have to be significantly different to allow a diagnosis, even with quantitative models. The overapproximations used to simplify the implementation of the set-observation lead to a long *delay* in proving the inconsistency of $f_1^{\approx}$, which is not significantly different from f_1. For an other I/O scenario, the two faults f_1 and $f_1^{\approx}$ could have remained candidates for the whole considered horizon.

For the same uncertainty assumptions, this can only be improved by means of a more precise observation. For example, by overapproximating the state-sets only every $M = 20$ loops, the set-observation has a higher computation cost but proves the inconsistency of $\mathcal{M}_{f_1^{\approx}}(\boldsymbol{p})$ sooner ($t_\perp = 1.10\,[\mathrm{s}]$). Using this diagnostic method a tradeoff is therefore to be found between the diagnostic accuracy (by means of the observer's accuracy) and the affordable computation burden.

A case of non-constant fault requires a relaxation of Assumption II which is pursued in [6]. By allowing the fault to vary over time, the guarantee to include the true fault in the set of fault candidates (as described in Def. 2.8) is lost. The induced set of faults can possibly become empty, which needs to be considered in the interpretation of the observation results for diagnosis (*cf.* Prop. 4.2).

6. Application to a cold rolling mill

This chapter describes the application of the guaranteed diagnostic method to a cold rolling mill. The focus lies on the detection of faults which disrupt the unwinding of the metal strip. This work is the result of a cooperation with ABB Corporate Research Germany and has been for its greater part published in [3, 7, 8]. In-depth details are found in the internal report [2].

6.1. Process description and diagnostic task

A case study of diagnosis for a rolling mill (Fig. 6.1) is presented. These machines are used in the metallurgy industry to flatten and/or reduce the thickness of long strips of metal. The strip is rolled onto a coiler, which may weigh many tons. In this process the metal is unwinded on one side of the mill, flows into the roll gap at very high speeds (up to 100 [km/h]) and is re-wrapped on the other side of the mill. Many constructions exist depending on the number of stands or on the number of rolls in each stand. For a single-stand rolling mill, this study focuses on the unwinding process and its possible faults.

Diagnosing faults during the unwinding is of foremost importance not only to assure a desired quality level of the end product but also because of tremendous costs associated with unplanned down-times. The study concentrates on four type of possible faults, $\mathbb{F} = \{f_0, f_1, f_2, f_3\}$ *cf.* Table 6.1. The available input-output measurements:

$$\boldsymbol{u} = (u_1, u_2)^T \quad \text{with} \quad u_1 = v_{laser}, \quad u_2 = F_t^{ref} \,, \tag{6.1}$$

$$\boldsymbol{y} = (y_1, y_2)^T \quad \text{with} \quad y_1 = \omega_{mot}, \quad y_2 = F_t \,. \tag{6.2}$$

This chapter only presents the fault detection task. For this purpose, only the model of the faultless behaviour $\mathcal{M}_{f_0}$ is used together with the diagnostic approach presented in Chapter 4 (Alg. IV). The faulty behaviours for f_0, f_1, f_2 and f_3 are realised in Dymola™ – a Modelica-based object-oriented modelling tool, [110] – such that sequences of I/O for all possible fault cases can be generated to test the diagnosis.

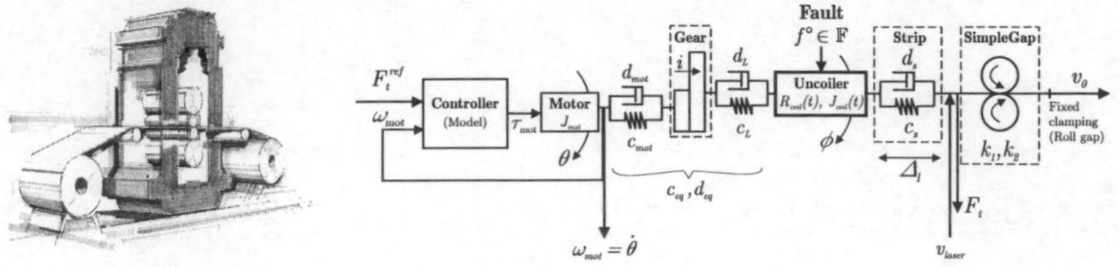

(a) Single-stand rolling mill (b) Object-oriented representation of the unwinding

Figure 6.1.: Rolling mill process example: unwinding a strip of metal into the roll gap.

Fault	Description
f_0	Faultless behaviour
f_1, f_2	Slipping faults
f_3	Sticking faults

Table 6.1.: Faults in the unwinding

6.2. Nonlinear unwinding model

6.2.1. Differential equations of the faultless behaviour

Two models of the unwinding are used for this study. The role of each model is explained below.

- The *diagnostic model* represents the linear discrete-time state-space model used in the diagnostic algorithm. This model is based on the available input and output measurements (6.1)–(6.2). In the presented study, only a model $\mathcal{M}_{f_0}$ of the faultless behaviour is used. The derivation of the diagnostic model is shown in Section 6.4.

- The *simulation model* (in Dymola™) is used to generate process signals corresponding to different working conditions of the unwinding, including the faulty behaviours. In this manner, parameter values, set-point trajectories (v_0 and F_t^{ref}) and occurring fault ($f^\circ \in \mathbb{F}$) are chosen at will. Such tests cannot be carried out in an actual metal processing factory.

  The simulation model is nonlinear continuous-time and is implemented in Dymola™. Its simulation generates the signals needed to pursue the diagnosis. This model is described in following.

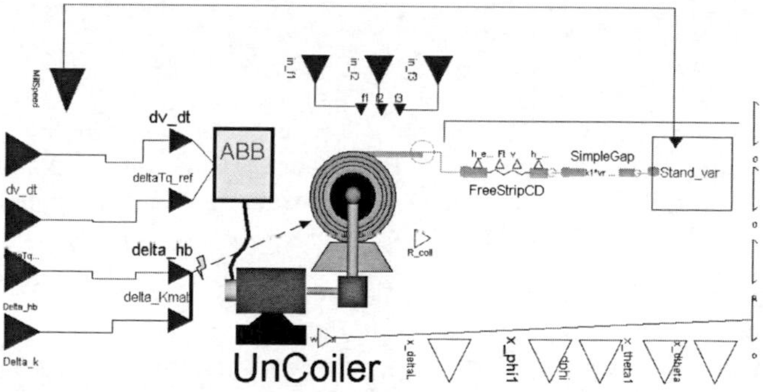

Figure 6.2.: Dymola model of the unwinding (screen-shot of upper hierarchical level)

A structural distinction between these models lies in the use of the set-point v_0 as an input to the simulation model. Therefrom, the simulation generates ω_{mot} and F_t (the available *output* measurements) as well as v_{laser} (an available *input* measurement). This setup is used as v_0 is a set-point signal which can be artificially chosen (*e.g.* a step input), whereas v_{laser} is a more complex signal resulting from the mill's dynamics.

The interaction of these models is shown later in Fig. 6.3: the simulation model is only used if I/O behaviours are generated in simulation, not if actual I/O measurements from the mill are used. The diagnosis always uses the same diagnostic model.

Nonlinear differential equations. The equations common to both the simulation and the diagnostic models are now derived. The object-oriented representation of the unwinding shown in Fig. 6.1(b) is realised in Dymola™, of which a screen-shot is shown in Fig. 6.2. For each component of the unwinding (controller, strip, uncoiler, ...), a dynamic model is obtained using ABB's internal knowledge and standard literature [24,50]. These equations are thoroughly described in [1] and were presented at [8]. Using the parameter and signals defined in Tables 6.2 and 6.3, the following differential equations are gathered:

$$J_{mot}\,\ddot{\theta} = \tau_{mot}(\omega_{mot}, F_t^{ref}) + c_{eq}(i\,\phi - \theta) + d_{eq}(i\,\dot{\phi} - \dot{\theta}) \tag{6.3a}$$

$$J_{coil}\,\ddot{\phi} = -\dot{J}_{coil}\,\dot{\phi} + F_t\,R_{coil} - i\,c_{eq}(i\,\phi - \theta) - i\,d_{eq}(i\,\dot{\phi} - \dot{\theta}) \tag{6.3b}$$

$$\dot{\Delta}_l = v_{laser} - R_{coil}\dot{\phi} \tag{6.3c}$$

$$\dot{J}_{coil} = -\rho\,W\,h_0\,R_{coil}^3\,\dot{\phi} \tag{6.3d}$$

$$\dot{R}_{coil} = -\frac{h_0}{2\pi}\,\dot{\phi} \tag{6.3e}$$

$$v_{laser} = v_0\big(k_1 + k_2(F_t - F_t^{ref})\big)\ . \tag{6.3f}$$

The torque signal $\tau_{mot}(\omega_{mot}, F_t^{ref})$ is computed by a model-based controller at hand of the measurement ω_{mot} and the set-point F_t^{ref}. Further details about this controller, as well as proper initialisation values for these differential equations, are found in the aforementioned internal report.

The structural distinction between the simulation and diagnostic models lies in the consideration, or not, of the simplified roll gap in Eq. (6.3f). In a simulation framework, v_0 (the strip speed set-point) is given and v_{laser} (the actual resulting strip speed) has to be determined using this equation. In the diagnostic framework, v_{laser} is measured and this equation becomes superfluous. This is advantageous for the diagnosis since k_1 and k_2 – which depend in part on the thickness reduction in the gap – are only roughly known, [35].

Parameter	Description	Unit
c_{eq}	Rotational spring constant (unwinding's drive)	[N m/rad]
d_{eq}	Rotational damping constant	[N m s/rad]
c_s	Translational spring constant (strip)	[N/m]
d_s	Translational damping constant	[N s/m]
J_{mot}	Inertia of unwinding's drive	[kg m^2]
R_{mnd}	Radius of the mandrel	[m]
J_{coil}	Inertia uncoiler	[kg m^2]
R_{coil}	Radius of uncoiler	[m]
i	Gear ratio from drive to mandrel	[1]
h_0	Strip thickness	[m]
W	Strip width	[m]
ρ	Strip material density	[kg/m^3]
k_1, k_2	Parameters approximating a simple roll gap	[1], [N^{-1}]

Table 6.2.: Parameters of the unwinding process

6.2.2. Simulation model

The financial cost related with the operation of a rolling mill makes it unsuitable – at this stage of the project – to test the diagnostic method against faults deliberately provoked

Signal	Description	Unit
θ	Angular rotation of drive	[rad]
ϕ	Angular rotation of uncoiler	[rad]
Δ_l	Strip elongation	[m]
F_t	Strip tension	[N]
F_t^{ref}	Set-point for strip tension	[N]
$\omega_{mot} = \dot{\theta}$	Angular drive velocity	[rad/s]
$v_{strip} = R_{coil}\dot{\phi}$	Strip speed at uncoiler	[m/s]
v_0	Strip speed in roll gap (*v_roll*)	[m/s]
v_{laser}	Strip speed before roll gap entry	[m/s]

Table 6.3.: Signals of the unwinding process

in the mill. Therefore, it is necessary to have a precise model of the unwinding which generates input and output signals corresponding to different scenarios, *e.g.* different process parameters, set-points, measurement disturbances or occurring faults.

Fig. 6.3(a) shows how to test the diagnosis using the simulation model. In the long run, the diagnosis needs to be tested using measurement signals from the plant as shown in Fig. 6.3(b). This test requires a very good fit between diagnostic model and the actual plant behaviour, even in the faultless case. In such a configuration, the diagnostic model has to be simulated to assure that measured and simulated outputs correspond: in the case of a faultless behaviour $\boldsymbol{y} \approx \tilde{\boldsymbol{y}}$ should hold (see dashed lines in upper block of the figure).

The mathematical description of simulation model for the faultless behaviour is given in the following. It is then extended in the next section to describe the considered fault cases. Defining the state, input and output variables to be

$$\boldsymbol{x} = (x_1, x_2, x_3, x_4, x_5, x_6, x_7)^T = (\theta, \dot{\theta}, \phi, \dot{\phi}, \Delta_l, J_{coil}, R_{coil})^T$$
$$\boldsymbol{u} = (u_1, u_2)^T = (v_0, F_t^{ref})^T$$
$$\boldsymbol{y} = (y_1, y_2)^T = (\omega_{mot}, F_t, v_{laser})^T,$$

the differential Eqs. (6.3) are rewritten as a nonlinear continuous-time state-space model:

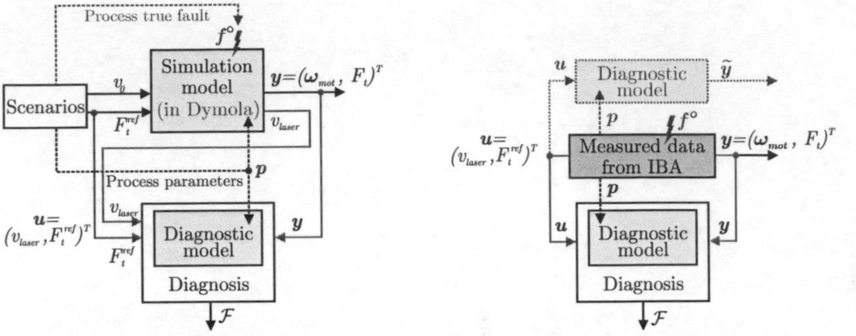

(a) Diagnosis using the simulation model (b) Diagnosis from measurements

Figure 6.3.: Two models used to test the unwinding diagnosis ("IBA" refers to a software used to analyse measurements of large automated facilities, [52]).

$$
\begin{aligned}
\dot{x}_1 &= x_2 \\
\dot{x}_2 &= \frac{1}{J_{mot}}\,\tau_{mot}(x_2, u_2) + \frac{c_{eq}}{J_{mot}}(i\,x_3 - x_1) + \frac{d_{eq}}{J_{mot}}(i\,x_4 - x_2) \\
\dot{x}_3 &= x_4 \\
\dot{x}_4 &= -\frac{\dot{x}_6}{x_6}\,x_4 + \frac{1}{x_6}\,F_t\,x_7 - \frac{i\,c_{eq}}{x_6}(i\,x_3 - x_1) - \frac{i\,d_{eq}}{x_6}(i\,x_4 - x_2) \\
\dot{x}_5 &= v_{laser} - x_7\,x_4 \\
\dot{x}_6 &= -\rho\,W\,h_0\,x_7^3\,x_4 \\
\dot{x}_7 &= -\frac{h_0}{2\pi}\,x_4
\end{aligned}
$$

with

$$
\begin{aligned}
v_{laser} &= u_1\big(k_1 + k_2(F_t - u_2)\big) \\
F_t &= c_s\,x_5 + d_s\,\dot{x}_5\ .
\end{aligned}
$$

The torque $\tau_{mot}(x_2, u_2)$ results from a model predictive controller, see Section 6.2.1. The simulation of this model generates the signals needed for diagnosis, *i.e.* the inputs v_{laser}, F_t^{ref} and the outputs ω_{mot}, F_t.

The above equations only describe the faultless behaviour of the unwinding. The mathematical model corresponding to the faulty behaviours is much more complex and not needed in this study. Hence, the next section only briefly describes how these fault behaviours have been realised in Dymola™. The resulting simulation model can generate I/O signals for any desired process scenario. Exemplary trajectories of the input and output signals generated using this model are shown in the figures concluding this chapter.

6.3. Faulty behaviours of the unwinding process

The study concentrates on faults which are *abrupt*, *non-periodic* and occur only for a *very short time*. Two classes of faults are considered: slipping faults and sticking faults. These faults all occur within the uncoiler component, as shown in Fig. 6.1(b). They represent an improper unwrapping of the metal strip from the uncoiler. In the following, a verbal description of the faults is given as well as an explanation of their implementation in the Dymola™ simulation model.

6.3.1. Description of faulty behaviours

Fault f_1: "mandrel slip." A slip occurs between the mandrel and the coil, caused by a bad connection between the first layers of the metal strip and the mandrel.

In the faultless case, the mandrel and the coil are modelled as a single and unified rotating component, described by the angle ϕ. In fault case f_1, the occurrence of the slip provokes a separation of the mandrel and the coil for a very short time: the mandrel spins freely, independently from the uncoiler's rotation. Therefore the slip requires to split the angle ϕ into ϕ_1 (mandrel's rotation) and ϕ_2 (uncoiler's rotation). During the fault occurrence, an increase of the state-space dimension occurs (ϕ and $\dot{\phi}$ are replaced by ϕ_1, ϕ_2, $\dot{\phi}_1$ and $\dot{\phi}_2$) and $v_{strip} = R_{coil}\dot{\phi}_2$.

A similar fault case – "coil slip" fault f_2 – was originally considered. This fault case describes a slip between two layers of metal within the uncoiler. As the simulated behaviour of fault f_2 is not distinguishable from that of fault f_1, the diagnostic results presented in Section 6.4.2 only consider f_1 as a possible slipping fault.

Fault f_3: "strip stick." In normal operation, the outermost layer of metal unwraps from the uncoiler and flows directly into the roll gap. In fault case f_3, the upper strip layer sticks to the coil surface for a short period of time. This sticking increases the strip tension F_t and the longer the fault lasts, the higher is the increase. The fault is said to be incipient, since its strength increases over time (or more specifically with the rotation ϕ of the uncoiler). The strip is then freed at once, causing large and dangerous oscillations in the process.

6.3.2. Implementation in Dymola^{tm}

The software model originally available only considers the faultless unwinding process. It is therefore extended such that any kind of fault behaviour can be generated, allowing a thorough test of the diagnosis (as in the configuration of Fig. 6.3(a)). At the end of this chapter, input-output trajectories for faulty behaviours are shown in Figs. 6.9 and 6.10, whereas a faultless behaviour is shown in Fig. 6.8.

Using an object-oriented modelling language, representing the faulty behaviours requires little mathematical description of the faults, as long as standard objects can be used.

Intuitively it is chosen to model the slipping faults with a *clutch* object and the sticking faults with a *brake* object, see Fig. 6.4.

Figure 6.4.: Dymola clutch object (to the left) and brake object (to the right)

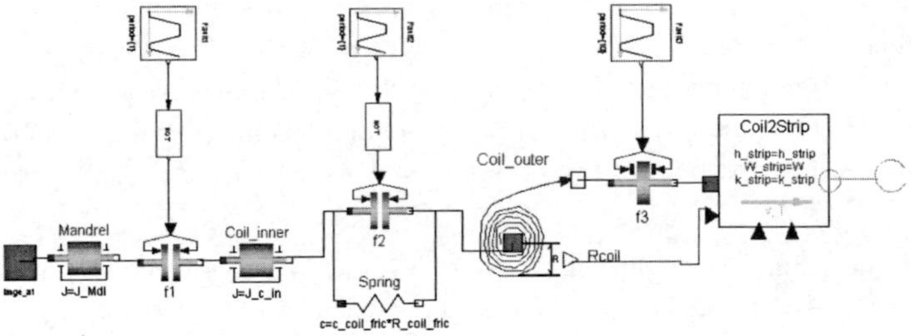

Figure 6.5.: Dymola modelling block of the faulty uncoiler (screen-shot)

In the faultless behaviour, the uncoiler is a rotating cylinder with time-varying radius and inertia. In the general case, the model shown in Fig. 6.5 extends the original model using the clutch and brake objects. From left to right the following objects are found.

- The "Flange" is a connector for rotational components. It connects the rotating shaft of the motor drive to the uncoiler.
- The mandrel is the metallic core of the uncoiler upon which the metal strip is wrapped It is modelled as a constant cylindrical inertia of radius R_{mnd}.
- The clutch named "f1" models the behaviour of fault f_1. The clutch causes a physical separation between the mandrel and the first layers of metal winded over the mandrel. This corresponds to the description of the fault which causes two masses to rotate with independent angles ϕ_1 (rotation of mandrel) and ϕ_2 (rotation of metal strip).
- The inertia "coil_inner" represents the first layers of winded strip between the mandrel (radius R_{mnd}) and the radius R_{coil_fric} at which the fault occurs. The latter radius locates the region where fault f_2 would occur.
- The clutch named "f2", together with the rotational spring in parallel, model the behaviour of fault f_2. The spring restrains the freedom of rotation of the two masses (whereas no physical component restrains the slipping during fault f_1). As fault f_2 is not considered in this thesis, this clutch remains closed at all times.
- The time-variable inertia "coil_outer" represents the outermost strip layers being unwinded of the uncoiler. These are the layers located between the fixed radius R_{coil_fric} and the time-varying radius $R_{coil}(t)$ of the uncoiler's surface.
- The brake named "f3" models the behaviour of fault f_3. This brake can slow down the rotation of the uncoiler and cause the increase in strip tension characteristic of this fault.
- The block "Coil2Strip" is used to transform the signals of the rotating movement into signals of a translating movement. It assures the compatibility with the next connectors in the model (not shown in the figure).
- The connector for strip output.

The signals triggering the faults are seen in the figure coming vertically into the clutches and the brake. The "NOT" blocks located at the entry of the two clutches are used to comply with the intuitive convention that a signal triggering faults is equal to zero when no fault occurs. It is not needed for the brake which intrinsically possesses this behaviour.

The new objects which were created to model the faults have been placed in *RMlib_FDI* library (*R*olling *M*ill *lib*rary for *F*ault *D*etection and *I*solation), which is built on the same structure as the "Rolling Mill Library" used internally at ABB.

6.4. Diagnosis of the unwinding process

6.4.1. Diagnostic model

To diagnose the unwinding using Algorithm IV, a linear state-space model $\mathcal{M}_{f_0}$ of the faultless behaviour is sought with inputs and outputs:

$$\begin{aligned} \boldsymbol{u} &= (u_1, u_2)^T \quad \text{with} \quad u_1 = v_{laser}, \quad u_2 = F_t^{ref} \\ \boldsymbol{y} &= (y_1, y_2)^T \quad \text{with} \quad y_1 = \omega_{mot}, \quad y_2 = F_t \ . \end{aligned}$$

The following describes the intermediary steps needed to transform the nonlinear model from Eqs. (6.3) into the desired form.

Nonlinear state-space model. The differential Eqs. (6.3) are rewritten using the state-space formalism for the above input and output specifications. The same state vector as for the simulation model is used

$$\boldsymbol{x} = (x_1, x_2, x_3, x_4, x_5, x_6, x_7)^T = (\theta, \dot{\theta}, \phi, \dot{\phi}, \Delta_l, J_{coil}, R_{coil})^T$$

resulting in the nonlinear state-space model:

$$\dot{x}_1 = x_2 \tag{6.5a}$$

$$\dot{x}_2 = \frac{1}{J_{mot}} \tau_{mot}(x_2, u_2) + \frac{c_{eq}}{J_{mot}}(i\, x_3 - x_1) + \frac{d_{eq}}{J_{mot}}(i\, x_4 - x_2) \tag{6.5b}$$

$$\dot{x}_3 = x_4 \tag{6.5c}$$

$$\dot{x}_4 = -\frac{\dot{x}_6}{x_6}\, x_4 + \frac{1}{x_6}\, F_t\, x_7 - \frac{i\, c_{eq}}{x_6}(i\, x_3 - x_1) - \frac{i\, d_{eq}}{x_6}(i\, x_4 - x_2) \tag{6.5d}$$

$$\dot{x}_5 = u_1 - x_7\, x_4 \tag{6.5e}$$

$$\dot{x}_6 = -\rho\, W\, h_0\, x_7^3\, x_4 \tag{6.5f}$$

$$\dot{x}_7 = -\frac{h_0}{2\pi}\, x_4 \tag{6.5g}$$

with

$$F_t = c_s\, x_5 + d_s\, \dot{x}_5 \ . \tag{6.5h}$$

Simplified linear model. Based on the analysis of simulated signals [1], it was made clear that the considered faults can only be detected in the short time lapse of their occurrence, and not later. Therefrom, a simplification of the nonlinear model is undergone, which assumes the radius of the uncoiler R_{coil} (respectively its moment of inertia J_{coil}) to vary negligibly within a short time window. Let such a time window begin at time t_0. A simplified model of the unwinding is obtained by replacing $x_7 = R_{coil}(t)$ with the constant $R_{coil}(t_0)$ (resp. x_6 with $J_{coil}(t_0)$) and by letting its derivative vanish such that $\dot{x}_7 = \dot{R}_{coil}(t) \approx 0$ (resp. $\dot{J}_{coil} \approx 0$).

The state vector reduces to

$$\boldsymbol{x} = (x_1, x_2, x_3, x_4, x_5)^T = \left(\theta, \dot{\theta}, \phi, \dot{\phi}, \Delta_l\right)^T .$$

By separating the model predictive controller signal into

$$\tau_{mot}(x_2, u_2) = \tau_{mot}^{x_2}\, x_2 + \tau_{mot}^{u_2}\, u_2 \ ,$$

a linear state-space model results:

$$\dot{x}_1 = x_2 \tag{6.6a}$$

$$\dot{x}_2 = \frac{\tau_{mot}^{x_2}}{J_{mot}}\, x_2 + \frac{\tau_{mot}^{u_2}}{J_{mot}}\, u_2 + \frac{c_{eq}}{J_{mot}}(i\, x_3 - x_1) + \frac{d_{eq}}{J_{mot}}(i\, x_4 - x_2) \tag{6.6b}$$

$$\dot{x}_3 = x_4 \tag{6.6c}$$

$$\dot{x}_4 = \frac{R_{coil}(t_0)}{J_{coil}(t_0)}\, F_t - \frac{i\, c_{eq}}{J_{coil}(t_0)}(i\, x_3 - x_1) - \frac{i\, d_{eq}}{J_{coil}(t_0)}(i\, x_4 - x_2) \tag{6.6d}$$

$$\dot{x}_5 = u_1 - R_{coil}(t_0)\, x_4 \tag{6.6e}$$

with

$$F_t :- F_t(x_5, u_1, x_4) = c_s\, x_5 + d_s\, \dot{x}_5 \tag{6.6f}$$

and the outputs

$$y_1 = \omega_{mot} = x_2 \tag{6.6g}$$

$$y_2 = F_t = c_s\, x_5 + d_s\, (u_1 - R_{coil}(t_0)\, x_4) \ . \tag{6.6h}$$

This model is linear but *structurally* unobservable. This is shown using a Kalman decomposition of the model, [78]. Operating a coordinate transformation such that

$$\boldsymbol{x} = (x_1, x_2, x_3, x_4, x_5)^T = \left(\dot{\theta}, (i\phi - \theta), \dot{\phi}, \Delta_l, (i\phi + \theta)\right)^T$$

the following model is obtained (using now $\tau_{mot}(x_1, u_2) = \tau_{mot}^{x_1} x_1 + \tau_{mot}^{u_2} u_2$)

$$\dot{x}_1 = \ddot{\theta} = \frac{\tau_{mot}^{x_1}}{J_{mot}} x_1 + \frac{\tau_{mot}^{u_2}}{J_{mot}} u_2 + \frac{c_{eq}}{J_{mot}} x_2 + \frac{d_{eq}}{J_{mot}} (i\, x_3 - x_1) \tag{6.7a}$$

$$\dot{x}_2 = i\dot{\phi} - \dot{\theta} = i\, x_3 - x_1 \tag{6.7b}$$

$$\dot{x}_3 = \ddot{\phi} = \frac{R_{coil}(t_0)}{J_{coil}(t_0)} F_t - \frac{i\, c_{eq}}{J_{coil}(t_0)} x_2 - \frac{i\, d_{eq}}{J_{coil}(t_0)} (i\, x_3 - x_1) \tag{6.7c}$$

$$\dot{x}_4 = u_1 - R_{coil}(t_0)\, x_3 \tag{6.7d}$$

$$\dot{x}_5 = i\dot{\phi} + \theta = i\, x_3 - x_1 \tag{6.7e}$$

with

$$F_t := F_t(x_4, u_1, x_3) = c_s\, x_4 + d_s\, \dot{x}_4 \tag{6.7f}$$

and the outputs

$$y_1 = \omega_{mot} = x_1 \tag{6.7g}$$

$$y_2 = F_t = F_t(x_4, u_1, x_3) = c_s\, x_4 + d_s\,(u_1 - R_{coil}(t_0)\, x_3)\ . \tag{6.7h}$$

The structural unobservability of the above model is illustrated in Fig. 6.6 which depicts the two signal-flow graphs corresponding to Eqs. (6.6) and (6.7). The structural analysis of the second graph shows that its state x_5 does not influence the outputs and hence cannot be observed, [76].

Diagnostic model. A model is then sought which can be used for state-set observation as needed for the diagnosis with Alg. IV. Such model is obtained by considering solely the *observable subspace* of the simplified linear model (6.7), which corresponds to the state vector

$$\boldsymbol{x} = (x_1, x_2, x_3, x_4)^T = \left(\dot{\theta}, (i\phi - \theta), \dot{\phi}, \Delta_l\right)^T .$$

Therefore, using the continuous-time state-space notation

$$\dot{\boldsymbol{x}} = \boldsymbol{A}_c \boldsymbol{x} + \boldsymbol{B}_c \boldsymbol{u}$$
$$\boldsymbol{y} = \boldsymbol{C}\boldsymbol{x} + \boldsymbol{D}\boldsymbol{u}\ ,$$

the observable model is:

$$\boldsymbol{A}_c = \begin{bmatrix} \frac{\tau_{mot}^{x_1} - d_{eq}}{J_{mot}} & \frac{c_{eq}}{J_{mot}} & \frac{i\, d_{eq}}{J_{mot}} & 0 \\ -1 & 0 & i & 0 \\ \frac{i\, d_{eq}}{J_{coil}(t_0)} & \frac{-i\, c_{eq}}{J_{coil}(t_0)} & \frac{-i^2 d_{eq} - R_{coil}(t_0)^2 d_s}{J_{coil}(t_0)} & c_s \frac{R_{coil}(t_0)}{J_{coil}(t_0)} \\ 0 & 0 & -R_{coil}(t_0) & 0 \end{bmatrix} \tag{6.8a}$$

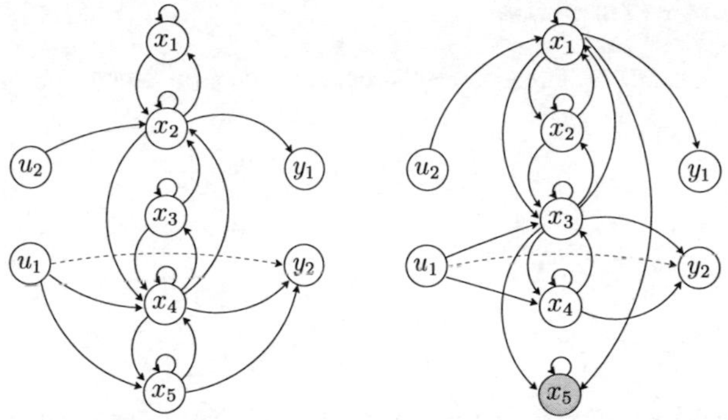

(a) Signal-flow graph corresponding to Model (6.6)

(b) Signal-flow graph corresponding to Model (6.7)

Figure 6.6.: Signal-flow graphs of two unwinding models (to increase readability the direct feedthrough is drawn dashed). On the right, no path leads from x_5 to an output, hence the state is unobservable.

$$\boldsymbol{B}_c = \begin{bmatrix} 0 & \frac{\tau_{mot}^{u_2}}{J_{mot}} \\ 0 & 0 \\ d_s \frac{R_{coil}(t_0)}{J_{coil}(0)} & 0 \\ 1 & 0 \end{bmatrix} \tag{6.8b}$$

$$\boldsymbol{C} = \begin{bmatrix} 1 & 0 & 0 & 0 \\ 0 & 0 & -d_s R_{coil}(t_0) & c_s \end{bmatrix} \tag{6.8c}$$

$$\boldsymbol{D} = \begin{bmatrix} 0 & 0 \\ d_s & 0 \end{bmatrix} . \tag{6.8d}$$

This model is converted into the discrete-time state-space model $\mathcal{M}_{f0}$

$$\begin{aligned} \boldsymbol{x}(k+1) &= \boldsymbol{A}_{f0}\boldsymbol{x}(k) + \boldsymbol{B}_{f0}\boldsymbol{u}(k) \\ \boldsymbol{y}(k) &= \boldsymbol{C}_{f0}\boldsymbol{x}(k) + \boldsymbol{D}_{f0}\boldsymbol{u}(k) \end{aligned}$$

using standard sampling methods. For example, using a Zero-Order-Hold with sampling rate T_s, results in the matrices [79]:

$$\boldsymbol{A}_{f0} = e^{\boldsymbol{A}_c T_s} \quad \text{and} \quad \boldsymbol{B}_{f0} = \int_0^{T_s} e^{\boldsymbol{A}_c \alpha} \boldsymbol{B}_c \mathrm{d}\alpha \, .$$

6.4.2. Results in simulation

This section presents fault detection results obtained with the model $\mathcal{M}_{f_0}$ of the faultless unwinding.

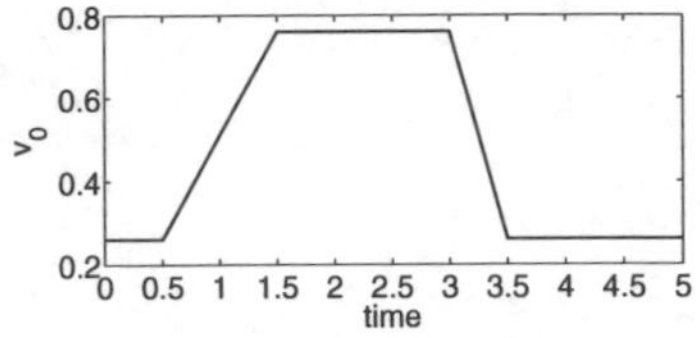

Figure 6.7.: Trajectory of the set-point v_0 used a input to the simulation model

Simulation of faultless and faulty behaviours. The I/O signals presented in this section are all obtained using the Dymola™ simulation model as illustrated in Fig. 6.3(a), page 100. The trajectory used for the set-point v_0 is shown in Fig. 6.7 and the process parameters correspond to a thick aluminium trip. This trapezoidal-shaped profile is typical for the operation of a rolling mill with speed-up, plateau and slow-down phases.

When simulating faulty behaviours, the fault (f_1 or f_3) is triggered at time $t = 1.3\,[\mathrm{s}]$.

Parameter setup for the diagnosis. The guaranteed diagnosis only requires to set appropriate bounds $\boldsymbol{e}_u$ and $\boldsymbol{e}_y$ describing the measurement uncertainties. For simplicity no input uncertainties are considered, *i.e.* $\boldsymbol{e}_u = (0,\ 0)^T$ and the input set is

$$\mathcal{U}(k) := \{\tilde{\boldsymbol{u}}(k)\} .$$

The output uncertainties are defined as a percentage of the measured output $\boldsymbol{y}(k) = (y_1(k),\ y_2(k))^T$. With 1% uncertainties on the first output $y_1(k)$ and 10% on the second output $y_2(k)$. The output set is

$$\mathcal{Y}(k) := \{\boldsymbol{y} \in \mathbb{R}^r \mid |\boldsymbol{y} - \tilde{\boldsymbol{y}}(k)| \leq |\tilde{\boldsymbol{y}}(k)| \cdot \mathrm{diag}(\tfrac{1}{100}, \tfrac{10}{100})\}$$

where

$$\mathrm{diag}(a_1, \ldots, a_r) = \begin{bmatrix} a_1 & & \mathbf{0} \\ & \ddots & \\ \mathbf{0} & & a_r \end{bmatrix} .$$

In the lower part of all Figs. 6.8 through 6.10, the state-sets $\mathcal{X}_{f_0}$ using the solely available model of the faultless unwinding $\mathcal{M}_{f_0}$ are shown. The sets are plotted using their projection along each state dimension, hence the intervals drawn for each state variable. The axes are named using the following short-hand notation for the four state variables of the diagnostic model :

- "dt" stands for $x_1 = \dot{\theta}$,
- "ip-t" stands for $x_2 = i\phi - \theta$,
- "dp" stands for $x_3 = \dot{\phi}$ and
- "Dl" stands for $x_4 = \Delta_l$.

Diagnostic results. As a single model is considered for the diagnosis, the bar diagrams used in Section 5.3 are not shown and only the state-set observation results are plotted. If the state-sets vanish such that $\mathcal{X}_{f_0}(\bar{k}) = \varnothing$, a fault is detected since $f_0 \notin \mathcal{F}(k)$, $\forall k \geq \bar{k}$.

The plots for the case of the faultless behaviour ($f^\circ = f_0$) are shown in Fig. 6.8. The state sets $\mathcal{X}_{f_0}$ are computed throughout the measurement horizon. Thus, the model $\mathcal{M}_{f_0}$ is consistent with the measurements, which is the desired result.

The plots for the faulty behaviour $f^\circ = f_1$ and $f^\circ = f_3$ are shown in Fig. 6.9 and Fig. 6.10. In both fault cases, the state-sets $\mathcal{X}_{f_0}$ vanish shortly after occurrence of the fault, causing their detection. Again, this demonstrates a successful diagnosis as it is recognised that the diagnostic model $\mathcal{M}_{f_0}$ (of the faultless behaviour) is inconsistent with the measured I/O behaviour.

More thorough tests of the diagnosis for different process parameters and diagnostic parameters are conducted in [2].

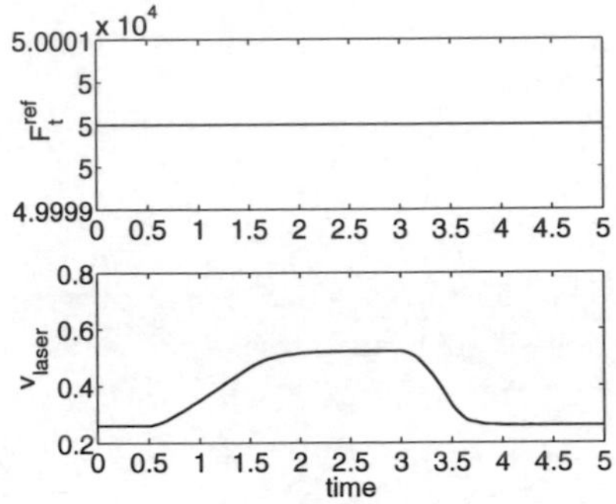

(a) Inputs

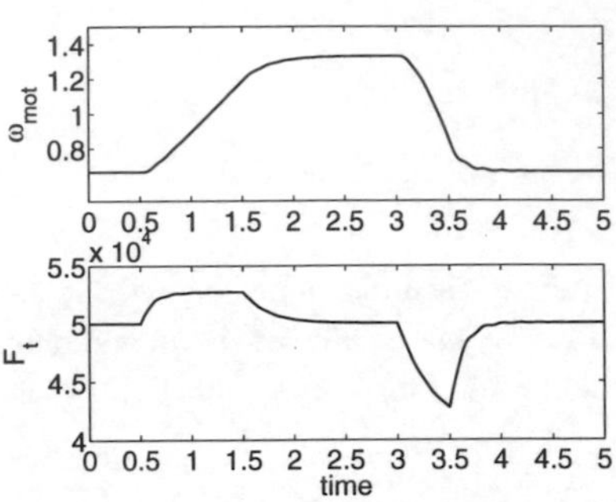

(b) Outputs

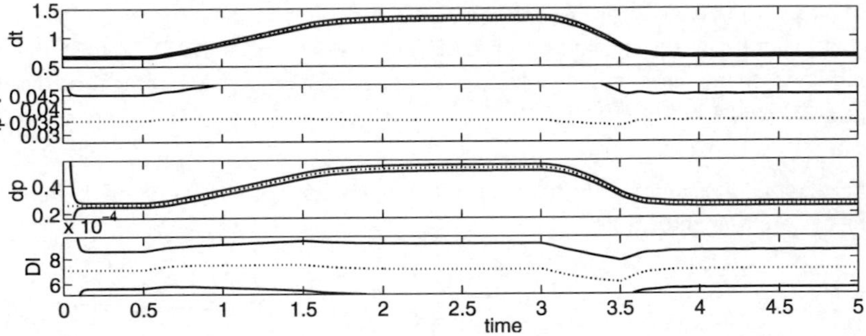

(c) Reconstructed state-sets $\mathcal{X}_{f_0}$

Figure 6.8.: Faultfree scenario $f^\circ = f_0$. The diagnostic model $\mathcal{M}_{f_0}$ is consistent with the I/O behaviour. No fault is detected which is the correct diagnostic result.

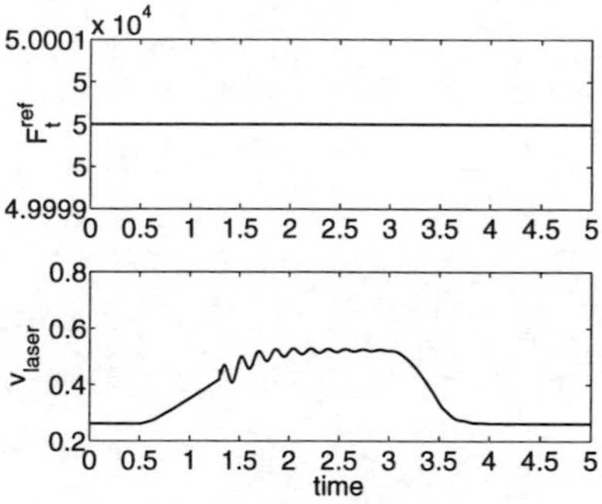

(a) Inputs

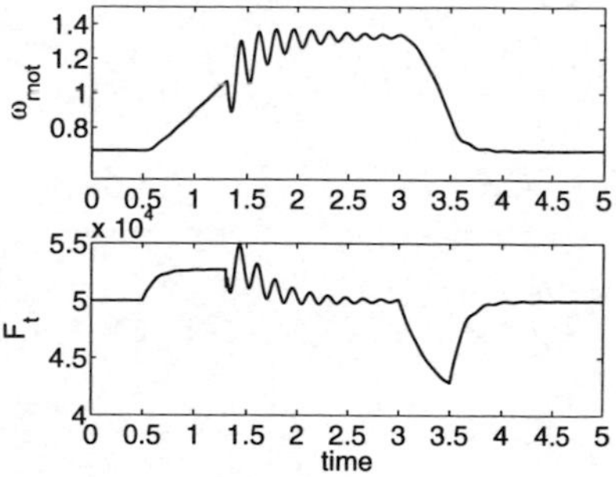

(b) Outputs

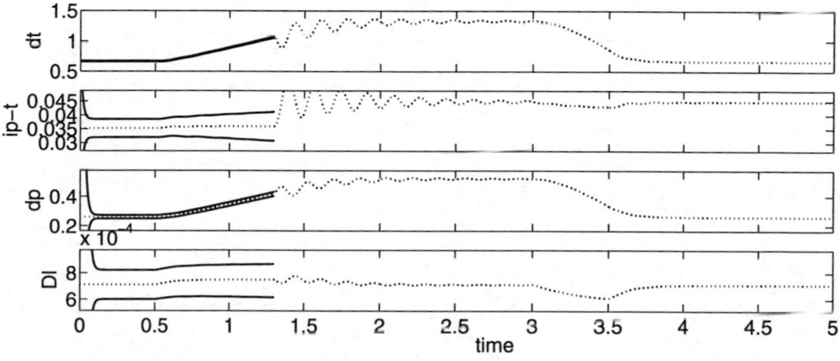

(c) Reconstructed state-sets $\mathcal{X}_{f_0}$

Figure 6.9.: Mandrel-slip fault scenario $f^\circ = f_1$. The consistency with the diagnostic model $\mathcal{M}_{f_0}$ is lost very soon after the fault occurs ($t = 1.3\,[\mathrm{s}]$), hence the fault is detected. (Very similar results are obtained for a coil-slip fault f_2).

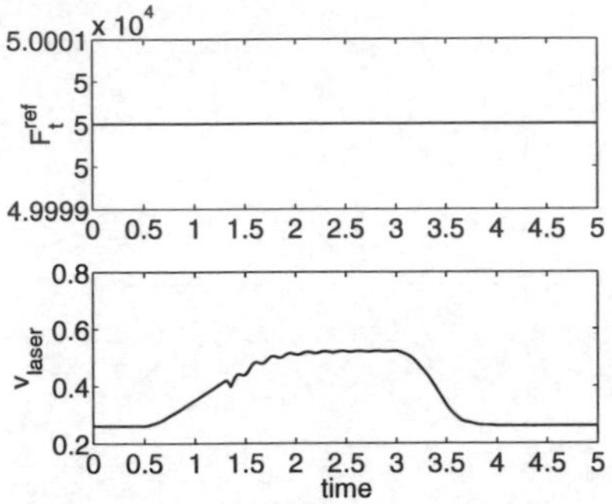

(a) Inputs

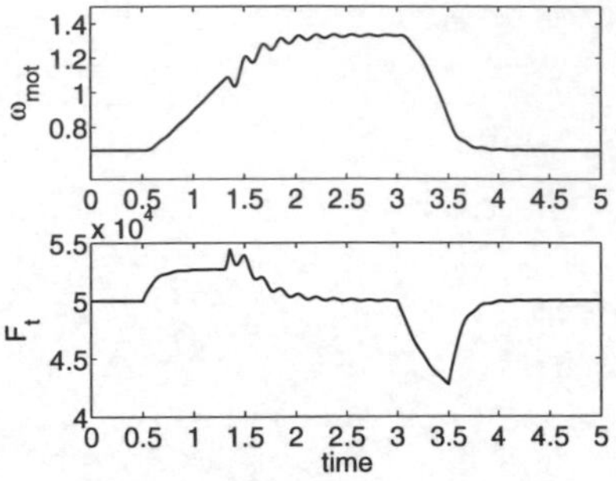

(b) Outputs

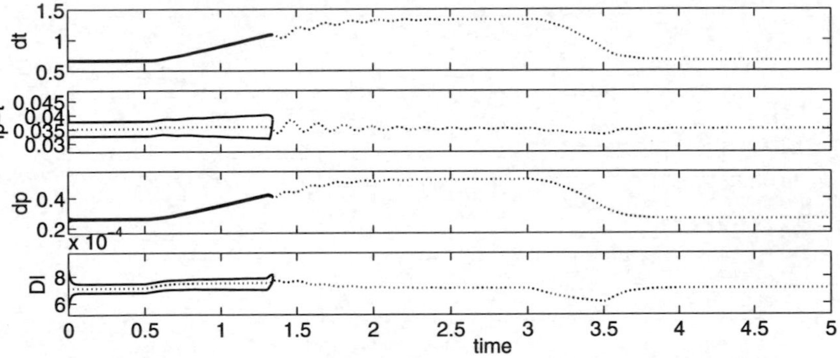

(c) Reconstructed state-sets $\mathcal{X}_{f_0}$

Figure 6.10.: Strip-stick fault scenario $f^\circ = f_3$. The consistency with the diagnostic model $\mathcal{M}_{f_0}$ is lost very soon after the fault occurs ($t = 1.3\,[\mathrm{s}]$), hence the fault is detected.

7. Summary and conclusions

Main results and contributions

This thesis contributes a novel approach in the class of observation-based diagnostic schemes. The proposed diagnostic method allows a simple and explicit treatment of uncertainties, both in the measurements and in the model of the process. The diagnosis is based on following information:

- The input and output of the process are disturbed in a non-stochastic manner. An unknown-but-bounded assumption is used to consider these uncertainties and results in a set-valued description of the input and output measurements.
- The process behaviour is represented using state-space models, such that each fault affecting the process has a distinct state-space model, which is a complete representation of the corresponding fault behaviour. This model may contain uncertain parameters in its system matrix.

Given the set-valued measurements, a state observer is designed which computes the set of all states possibly reached for a given model. If the model used for the observation matches the measured input-output behaviour, this set of states is non-empty. The corresponding model is then said to be *consistent* with the measurements.

Repeating this verification for each fault model induces a set of all consistent fault models. This set is proven to contain the true fault and the diagnosis is said to be *guaranteed*.

The set-valued observer underlying the diagnosis is implemented using a polyhedric description of the state-sets. This formalism is preferred to others (*e.g.* using ellipsoids) as it allows the computation of the *minimal set* of states. This is the smallest set containing exactly all states possibly reached for a given model. Deriving the minimal set is computationally expensive, but yields the most precise diagnostic result (Theorem 10).

It is shown that the minimal set cannot be described for an uncertain model as this necessitates the description of non-convex sets. Therefore an overapproximation of the minimal set is computed instead. This choice allows to implement a faster observation algorithm and most of all assures that the diagnostic result contains the true fault (since the diagnosis uses this overapproximated observation to verify model consistencies).

Compared to other diagnostic schemes the proposed method does not require an adjustment of its sensitivity to uncertainties. The robustness is acquired from the explicit consideration of the uncertainties within the set-observer and hence avoids the costly analysis of numerous simulations for different uncertainty assumptions.

The diagnosis is illustrated using a second-order motor example, which demonstrates the guaranteed diagnostic result obtained despite measurement and model uncertainties. Furthermore, a case study of an industrial application is shown in which the proposed method is used for fault detection.

Outlook

Future work related to this thesis is twofold and concerns the diagnosis on the one hand and the set-valued observation on the other hand. By construction, improvements to the latter also improve the former.

With respect to **diagnosis** determining criteria and properties of the *system diagnosability* remains to be studied. The proposed method is shown to be robust towards uncertainties in the measurements and in the models. However, too generous uncertainty bounds assuredly leads to the undetectability of the faults. Conditions for the diagnosability of a fault, which then depends on the amplitude of the uncertainties, are sought.

The present method focuses on a single diagnostic principle which verifies the consistency of given models. Other principles can be used to determine *distinguishing fault signatures.* In [2] for example, once a fault is detected (using only the model of the faultless behaviour), the duration of the inconsistency is determined and a pattern is sought which reestablishes the consistency. If the duration and/or pattern is characteristic of one of the faults, the diagnosis is improved without having used a state-space model to represent the faulty behaviour. Such techniques can refine the diagnosis, especially in the case of faults with short durations.

With respect to **set-valued observation** the following paragraphs lists several open research topics.

In the present approach the uncertain parameters are time-varying within given bounds. Considering *uncertain but constant parameters* is of high practical relevance. Approaches for this already exist in the area of controller design, *e.g.* [68], but integrating these in the present algorithm require presumably substantial and structural modifications.

The *decomposition of the system matrix* undertaken in Eq. (2.10) is not unique and its choice influences the observation of uncertain models in two respects (Prop. 3.4): First by modifying the computed predicted set (which depends on the matrix $\boldsymbol{A}_{f,0}$), and second by possibly easing the solution of the optimisation problem (*e.g.* if the parameters are positive, *i.e.* $\mathbb{P} \subset \mathbb{R}_+^{N_p}$). Finding a decomposition favourable to the diagnosis is to be studied.

The extension to *uncertain parameters in any matrix* of the state-space model would result in a more general algorithm. The case of an uncertain output matrix is especially non-trivial as this causes hyperboloidal measured state-sets (non-convex sets). An approach based on ellipsoidal sets is proposed in [91].

Finally, *nonlinear state-space* models, or a specific class thereof, could be considered. If the outputs are further assumed to be a linear combination of the states, this solely implies a modification in the computation of the predicted set such that existing results in the area of set-valued simulation can be used. Cases handling nonlinear functions are considered for example in [26, 49, 57]. A larger modification deals with the use of continuous-time dynamics, as for example in [72].

Appendix

A. Non-minimal set-observation case

This following shows that the recursive Eq. (3.9) given in Prop. 3.1 cannot be minimal for a simple system with direct feedthrough, *i.e.* $\mathcal{X}_f(k) \supsetneq \mathcal{X}_f^*(k \mid 0 \dots k)$.

Consider the first-order dynamical model $\mathcal{M}_f$

$$\begin{aligned} x(k+1) &= 0.8x(k) + u(k), \\ y(k) &= x(k) + u(k). \end{aligned}$$

For the true initial state $x^\circ(0) = 0.8$ and the true input sequence

$$U^\circ(0 \dots 1) = (0, 0) \; ,$$

the model generates the state and output sequences

$$\begin{aligned} X^\circ(0 \dots 1) &= (0.80, 0.64) \; , \\ Y^\circ(0 \dots 1) &= (0.80, 0.64) \; . \end{aligned}$$

Since this is a one-dimensional case, all considered sets are intervals. For a certain assumption on the measurement error, let the sequence of I/O sets be

$$\begin{aligned} \mathcal{U}(0 \dots 1) &= \big([0.0,\ 0.2],\ [0.0,\ 0.2]\big), \\ \mathcal{Y}(0 \dots 1) &= \big([-1.0,\ 1.0],\ [-0.1,\ 1.0]\big), \end{aligned}$$

such that at all times $k = 0$ and $k = 1$ the true I/O sequence is admissible: $u^\circ(k) \in \mathcal{U}(k)$ and $y^\circ(k) \in \mathcal{Y}(k)$.

Recursive description. At first the sequence of state-sets $\mathcal{X}_f(0 \dots 1) = (\mathcal{X}_f(0), \mathcal{X}_f(1))$ obtained using the recursion from Eq. (3.9) is computed for this example:

$$\begin{aligned} \mathcal{X}_f(k) = \big\{ x \in \mathbb{R} \mid x = (0.8\hat{x} + \hat{u}) \text{ with } \hat{x} \in \mathcal{X}_f(k-1),\ \hat{u} \in \mathcal{U}(k-1) \\ \text{and } (x + \check{u}) \in \mathcal{Y}(k) \text{ with } \check{u} \in \mathcal{U}(k) \big\} \; . \end{aligned}$$

The sequence of sets is initialised with $\mathcal{X}(-1) = \mathbb{R}^1 = \mathbb{R}$.

For $k = 0$, there is no knowledge about the input set $\mathcal{U}(-1)$, hence the prediction is left out and the set is computed as

$$\mathcal{X}_f(0) = \{x \in \mathbb{R} \mid (x + \check{u}) \in \mathcal{Y}(0) \text{ with } \check{u} \in \mathcal{U}(0)\}$$
$$\mathcal{X}_f(0) = \{x \in \mathbb{R} \mid (x + \check{u}) \in [-1.0,\ 1.0] \text{ with } \check{u} \in [0.0,\ 0.2]\}$$
$$\text{hence } \mathcal{X}_f(0) = [-1.2,\ 1.0]\ .$$

For $k = 1$ the set is obtained as

$$\mathcal{X}_f(1) = \{x \in \mathbb{R} \mid x = (0.8\hat{x} + \hat{u}) \text{ with } \hat{x} \in \mathcal{X}_f(0),\ \hat{u} \in \mathcal{U}(0) \text{ and } (x + \check{u}) \in \mathcal{Y}(1) \text{ with } \check{u} \in \mathcal{U}(1)\}$$
$$\mathcal{X}_f(1) = \{x \in \mathbb{R} \mid x = (0.8\hat{x} + \hat{u}) \text{ with } \hat{x} \in [-1.2,\ 1.0],\ \hat{u} \in [0.0,\ 0.2], \text{ and } (x + \check{u}) \in [-0.1,\ 1.0] \text{ with } \check{u} \in [0.0,\ 0.2]\}$$
$$\mathcal{X}_f(1) = \{x \in \mathbb{R} \mid x \in [-0.96,\ 1.00] \text{ and } x \in [-0.3,\ 1.0]\}$$
$$\mathcal{X}_f(1) = [-0.96,\ 1.00] \cap [-0.3,\ 1.0]$$
$$\mathcal{X}_f(1) = [-0.3,\ 1.0]\ .$$

Minimal description. The minimal set of states is now computed. For the considered time horizon the set is recalled to be:

$$\mathcal{X}_f^*(1 \mid 0 \ldots 1) = \{x(1) \in \mathbb{R} \mid \exists x(0) \in \mathbb{R},\ \exists u(0) \in \mathcal{U}(0),\ \exists u(1) \in \mathcal{U}(1) \text{ such that } x(1) := \psi_{x,f}(x(0), u(0)) \text{ and } \Psi_{y,f}\big(x(0), (u(0), u(1))\big) \in \mathcal{Y}(0 \ldots 1)\}\ .$$

In other words, for all $x(1) \in \mathcal{X}_f^*(1 \mid 0 \ldots 1)$, there exists $x(0) \in \mathbb{R}$, $u(0) \in \mathcal{U}(0)$ and $u(1) \in \mathcal{U}(1)$ such that

$$x(1) := \psi_{x,f}(x(0), u(0)) = 0.8\,x(0) + u(0)\ , \tag{A.1}$$
$$\text{with } \psi_{y,f}(x(0), u(0)) = (x(0) + u(0)) \in \mathcal{Y}(0)\ , \tag{A.2}$$
$$\text{and } \psi_{y,f}(x(0), (u(0), u(1))) = \big(0.8\,x(1) + u(1)\big) \in \mathcal{Y}(1)\ . \tag{A.3}$$

Contradiction. It is now shown that the state $\bar{x}(1) = 1.0$ – which belongs to $\mathcal{X}_f(1)$ – does not belong to the minimal set of states $\mathcal{X}_f^*(1 \mid 0 \ldots 1)$, hence $\mathcal{X}_f(1) \neq \mathcal{X}_f^*(1 \mid 0 \ldots 1)$. The proof is done by contradiction.

Assume $\bar{x}(1) = 1.0 \in \mathcal{X}_f^*(1 \mid 0 \ldots 1)$, then Eqs. (A.1)–(A.3) must have a solution. As of Eq. (A.1) it is known that

$$\bar{x}(1) = 0.8\bar{x}(0) + \bar{u}(0) = 1.0$$

hence $1.0 \leq 0.8\bar{x}(0) + \bar{u}(0) \leq 1.0$.

Since $\mathcal{U}(0) = [0,\ 0.2]$, the input is limited to $0 \leq \bar{u}(0) \leq 0.2$ and therefrom

$$1.0 \leq \bar{x}(0) \leq 1.25 \ . \tag{A.4}$$

Using Eq. (A.2) with $\mathcal{Y}(0) = [-1.0,\ 1.0]$ leads to

$$-1.0 \leq \bar{x}(0) + \bar{u}(0) \leq 1.0 \tag{A.5}$$

The set of inequalities found in (A.4) and (A.5) have a *single* solution

$$\bar{x}(0) = 1.0$$
$$\bar{u}(0) = 0.0 \ .$$

However, according to Eq. (A.1), this combination of state and input leads to the prediction

$$0.8\bar{x}(0) + \bar{u}(0) = 0.8 \ \neq \ \bar{x}(1) = 1.0 \ .$$

Therefore the original assumption was false, *i.e.* $\bar{x}(1) \notin \mathcal{X}_f^*(1 \mid 0 \ldots 1)$, which finishes the proof.

This example only shows that the recursion from Eq. (3.9) is not minimal for models with direct feedthrough. The completeness of the recursion, however, is proven in Prop. 3.1, such that $\mathcal{X}_f(1) \supset \mathcal{X}_f^*(1 \mid 0 \ldots 1)$. The guaranteed observation is clear for this example since

$$x^\circ(0) = 0.80 \in \mathcal{X}_f(0) = [-1.2,\ 1.0] \ ,$$
$$x^\circ(1) = 0.64 \in \mathcal{X}_f(1) = [-0.3,\ 1.0] \ .$$

B. MATLAB™ so4cd: a toolbox for set-observation based diagnosis

The functions and scripts necessary to realise the state-set observation – both complete as of Alg. II and minimal as of Alg. III – as well as the consistency-based diagnosis from Alg. IV have been implemented in MATLAB™ and lumped into a toolbox named SO4CD (Set-Observation for Complete Diagnosis).

More information about MATLAB™ and Simulink™ is found in [83, 84]. The toolbox is known to work under both Windows™ and Linux operating systems.

Installation of toolbox

The toolbox consists of a directory "so4cd/" containing the following files and directories:

```
Contents.m
DEMO/
devel/
FigureTools/
intervals/
misc/
ResultTools/
SignalTools/
so4cd_lib/
so4cd_main/
so4cd_path.m
ToolboxTools/
```

The toolbox is installed by calling `so4cd_path` which adds the appropriate directories to the command path. Basic help is obtained with `help so4cd`.

Main functions

This section describes the main functions of the toolbox, many more are found and documented using the built-in help command.

>> `SO4CD_commandline` is the principal function of the SO4CD toolbox. It computes simultaneously the state-set observation and the diagnosis. Its syntax is

```
[ polytopes, consistencies ] = ...
    SO4CD_commandline(models,X0,inputs,outputs,bounding,options)
```

Its input arguments are defined as:

- `models` is a cell containing the discrete-time state-space models $\mathcal{M}_f$, $f \in \mathbb{F}$.
- `X0` is the a priori state-set $\mathcal{X}(0\,|\,{-1})$. To consider the entire state-space use the empty matrix: `X0=[]`. Otherwise use a n-by-2 matrix describing an interval set `X0=[x1_min, x1_max; x2_min, x2_max; ...]`.
- `inputs` is a structures-with-time containing the input sequence $\boldsymbol{U}(0\ldots\bar{k})$. This structure is obtained for example using the "Save to Workspace" block in Simulink™, which generates the needed fields `inputs.time` and `inputs.signals.values`.

  The conversion to a set-valued input sequence occurs internally using the fields `inputs.uncertainty.absolute` and `inputs.uncertainty.percentage` which have to be additionally defined.
- `outputs` is the output sequence $\boldsymbol{Y}(0\ldots\bar{k})$. It is defined just as `inputs`.
- `bounding` is a structure describing the type of overapproximation. An empty value (`bounding=[]`) defaults to an axis-parallel overapproximation. The full syntax is to use `bounding.method='e'` for a minimal set-observation (exact) and `bounding.method='a'` for an axis-parallel overapproximation (axis). Any other constraint orientation can be chosen with `bounding.method='f'` (free), which then requires `bounding.directions` to be the μ-by-n matrix describing the desired μ orientation vectors.
- `options` (optional argument) is a structure used to trigger specific behaviours of the function. The list of fields accepted in this structure are found with `help SO4CD_0_check_options`.

The output arguments are:

- `polytopes` is a cell-array with as many cells as they are models defined in `models`. Each cell contains a further cell-array representing the sequence of polytopes (state-sets) reconstructed for the considered model.

- `consistencies` is analogous to `polytopes` but describes exclusively the consistency result (boolean) for each given model over time. Hence it is a cell-array with as many cells as they are models in `models`. Each cell contains a row-vector representing a sequence of the consistency of the considered model with the input-output: so `consistencies{i}(k)==1` if the model `models{i}` is consistent up to time `k`, *i.e.* if `polytopes{i}{k}` is not empty. Conversely for `consistencies{i}(k)==0`.

Both output arguments are suitably drawn using the functions `func_result_setobs` and `func_result_diagnosis` described below.

>> `SO4CD_sfunc` is the S-function version of `SO4CD_commandline`. It is used to achieve a state-set observation and/or a diagnosis within Simulink™. Two blocks are found as in Fig. B.1. One block uses a dialog mask while the other has no mask and requires the user to give the arguments similarly as with `SO4CD_commandline`. Using the S-function, the real-valued I/O are lead into the block while the `inputs.uncertainty` and `outputs.uncertainty` fields are given as S-function parameters, see Fig. B.2.

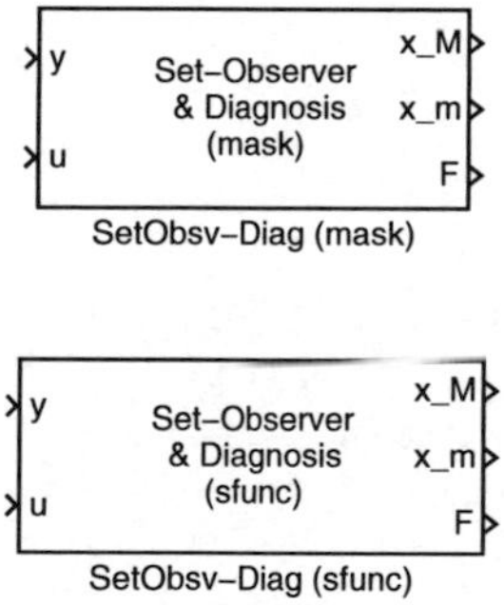

Figure B.1.: Content of the SO4CD Simulink™ library

>> `func_result_diagnosis` draws the bar diagrams used to represent the consistency of the models graphically (as in Fig. 5.13, page 90). It takes as mandatory argument the second output of `SO4CD_commandline`:

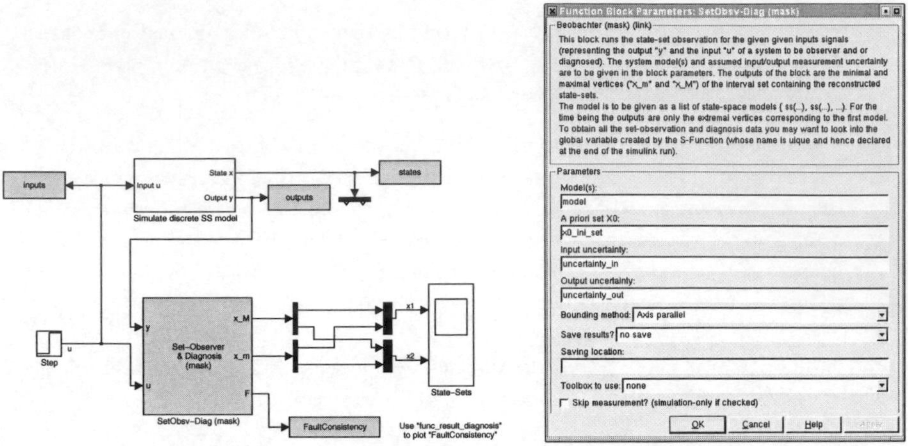

(a) Demo example in the toolbox (b) Masked version of the S-function

Figure B.2.: Using the SO4CD toolbox in Simulink™

```
func_result_diagnosis(consistencies)
```

An optional argument specifies the sample time T_s such that the bar diagram display continuous times $t = k\,T_s$ instead of discrete times k.

>> `func_result_setobs` draws the result of the state-set observation (as in Fig. 5.12(d), page 90). It takes as mandatory argument the first output of `SO4CD_commandline`:

```
func_result_setobs(polytopes)
```

An optional argument specifies a structure-with-time containing a state sequence to compare with (*e.g.* the true state sequence $\boldsymbol{X}^{\circ}(0\dots\bar{k})$ if it is known).

>> `func_multiplot` plots the multiple signals coexisting in a single structure-with-time in separate subplots. This is called for example as `func_multiplot(inputs)`.

C. Nomenclature

Symbols

The symbols used obey the following conventions:

- **scalars** are represented by italic letters (like n, i, t, e_{y_i}),
- **vectors** are represented by bold italic lower-case letters (like $\boldsymbol{x}$, $\boldsymbol{y}$, $\boldsymbol{c}_{f,i}^T$),
- **sequences of vectors** and **matrices** are represented by bold italic upper-case letters (like $\boldsymbol{U}$, $\boldsymbol{Y}$ and $\boldsymbol{A}$, $\boldsymbol{C}$),
- **sets** and **sequences of sets** are represented by calligraphic upper-case letters (like $\mathcal{X}$, $\mathcal{Y}$, $\mathcal{F}$).

The symbols frequently used in this thesis are listed below.

Symbol	Signification	Reference
$\models$	Symbol of consistency	Def. 2.6
$\not\models$, $\perp$	Symbols of inconsistency	
Γ_u	Input set generator function $\mathcal{U} = \Gamma_u(\boldsymbol{u})$	Eq. (2.15)
Γ_y	Output set generator function $\mathcal{Y} = \Gamma_y(\boldsymbol{y})$	Eq. (2.16)
$\mathcal{M}_f$	Model of the behaviour of fault f (state-space model)	Section 2.2
$\mathcal{M}_f(\boldsymbol{p})$	Exact model (with known parameter $\boldsymbol{p}$)	Eq. (2.2)
$\mathcal{M}_f(\mathbb{P})$	Uncertain model (with parameter $\boldsymbol{p} \in \mathbb{P}$)	Eq. (2.7)
$\psi_{x,f}$, $\Psi_{x,f}$	Input-to-state operators for model $\mathcal{M}_f$	Section 1.5
$\psi_{y,f}$, $\Psi y, f$	Input-to-output operators for model $\mathcal{M}_f$	Section 1.5
$F_{f,p}$, $F_{f,m}$, $F_\cap$, $F_\approx$	Set transformation operators	Alg. I
$F_\bullet$, $F_\bullet^*$	Any of the (minimal) set transformation operators	p. 41
$\mathbb{F}$	Finite set of considered faults, $\mathbb{F} = \{f_0, f_1, \ldots, f_N\}$	
$\mathbb{N}$	Set of nonnegative integers	

Symbol	Signification	Reference
$\mathbb{P}$	Set in which parameter vary $\boldsymbol{p} \in \mathbb{P} \subset \mathbb{R}^{N_p}$	Eq. (2.11)
$\mathbb{R}$	Set of real numbers	
$\mathbb{R}^n$	Set of real vectors of dimension n	
$2^{\mathbb{R}^n}$	Set of all subsets of $\mathbb{R}^n$ (power set)	
$\mathbb{R}^n_+$	Set of non-negative real vectors of dimension n	
$\mathbb{R}^{n \times m}$	Set of real matrices, with n rows and m columns	
$\mathbb{U}$	Universum	Def. 2.1
$\boldsymbol{A}_f$	System matrix of model $\mathcal{M}_f$,	
$\boldsymbol{B}_f$	Input matrix of model $\mathcal{M}_f$,	
$\boldsymbol{C}_f$	Output matrix of model $\mathcal{M}_f$,	
$\boldsymbol{D}_f$	Direct feedthrough matrix of model $\mathcal{M}_f$,	
$\boldsymbol{I}$, $\boldsymbol{I}_n$	Identity matrix (of dimension n)	
$\boldsymbol{S}_o$, $\boldsymbol{S}_r$	Observability and constructibility matrices	Theorem 6
$\boldsymbol{p}$	Parameter vector $\boldsymbol{p} = (p_1, p_2, \ldots, p_{N_p})^T$	Eq. (2.10)
$\boldsymbol{u}$	Input vector $\boldsymbol{u} = (u_1, u_2, \ldots, u_m)^T$	
$\boldsymbol{x}$	State vector $\boldsymbol{x} = (x_1, x_2, \ldots, x_n)^T$	
$\boldsymbol{x}_0$	Initial state $\boldsymbol{x}_0 = \boldsymbol{x}(0)$	
$\boldsymbol{y}$	Output vector $\boldsymbol{y} = (y_1, y_2, \ldots, y_r)^T$	
$\boldsymbol{c}^T_{f,j}$	j^{th} row of output matrix $\boldsymbol{C}_f$	
$\boldsymbol{d}^T_{f,j}$	j^{th} row of feedthrough matrix $\boldsymbol{D}_f$	
N	Number of considered faults	
N_p	Number of uncertain parameters	Eq. (2.10)
f°	True fault affecting a process, $f^\circ \in \mathbb{F}$	
n	Dimension of state-space	
m	Dimension of input-space	
r	Dimension of output-space	
u_i	i^{th} component of input vector $\boldsymbol{u}$	
x_i	i^{th} component of state vector $\boldsymbol{x}$	
y_i	i^{th} component of output vector $\boldsymbol{y}$	

Symbol	Signification	Reference
$\mathcal{B}^{\circ}$, $\tilde{\mathcal{B}}$	True behaviour and modelled behaviour	Section 2.1
$\mathcal{F}$	Set of fault vectors $\mathcal{F} \subset \mathbb{F}$	Def. 4.1
$\mathcal{F}^{\circ}$	True set of fault candidates	Def. 2.2
$\mathcal{F}^{*}$	Set of fault candidates	Def. 2.8
$\mathcal{U}$	Set of admissible input vectors $\boldsymbol{u} \in \mathcal{U} \subset \mathbb{R}^m$	
$\mathcal{X}$	Set of state vectors $\boldsymbol{x} \in \mathcal{X} \subset \mathbb{R}^n$	
$\mathcal{X}_f$	Set of states obtained for model $\mathcal{M}_f$	
$\mathcal{X}_f^*$	Minimal Set of states obtained for model $\mathcal{M}_f$	Section 3.1
$\mathcal{X}(0 \,\vert\, -1)$	A priori state-set	Alg. I
$\mathcal{X}_{f,p}$	Predicted state set	
$\mathcal{X}_{f,m}$	Measured state set	
$\mathcal{X}_{f,\cap}$	Corrected (or intersected) state set	
$\mathcal{X}_{f,\bullet}$	Any of $\mathcal{X}_{f,p}$, $\mathcal{X}_{f,m}$, $\mathcal{X}_{f,\cap}$	
$\mathcal{Y}$	Set of admissible outputs vectors $\boldsymbol{y} \in \mathcal{Y} \subset \mathbb{R}^r$	

Superscripts

The *true value* of a signal is indicated by the "o"-superscript. This usually refers to an unknown value, such as the true fault f° or the true (undisturbed) input signal $\boldsymbol{u}^{\circ}$.

Values which correspond to an *optimal result* are denoted by the "$*$"-superscript. Example of which are the minimal state-set observation result $\mathcal{X}^*$ (Section 3.1) or the set of fault candidates $\mathcal{F}^*$ (Def. 2.8).

Abbreviations

Acronym	Meaning	Reference
FDI	Fault Detection and Identification	
GBT	Geometrical Bounding Toolbox	[73]
IBA	IngenieurBüro Anhaus	[52], Chap. 6
I/O	Input-Output	
LMI	Linear Matrix Inequality	
LP	Linear Program	[22]
SSO	State-Set Observation	
s.t.	"such that" (mathematical notation)	

Bibliography

I. Author related references

[1] Lunze, J. and P. Planchon. *Modelling and Diagnosability Study of an Uncoiler.* Technical report, ABB DECRC Ladenburg, October 2002.

[2] Lunze, J. and P. Planchon. *Development and testing of process supervision methods using semi-quantitative models for cold rolling mills.* Technical report, ABB DECRC Ladenburg, October 2004. Final report of ABB-ATP cooperation.

[3] Lunze, J., P. Planchon and M. Rode. *Verfahren zur Fehlererkennung in industriellen Prozessen mit Hilfe von Intervallbeobachtern.* Patent DE102004025574, January 2006.

[4] Mleczko, M. *Development and Implementation of Algorithms for State-Set Observation.* Studienarbeit, Ruhr-Universität Bochum, Lehrstuhl für Automatisierungstechnik und Prozessinformatik, 2004.

[5] Planchon, P. *Polytopic state-set observation: Completeness and application to diagnosis.* Forschungsbericht 2006.02, Ruhr-Universität Bochum, Lehrstuhl für Automatisierungstechnik und Prozessinformatik, Bochum, February 2006.

[6] Planchon, P. and J. Lunze. Robust diagnosis using state-set observation. In *Proceedings of SAFEPROCESS*, Beijing, China, 2006.

[7] Planchon, P., J. Lunze and M. Rode. Diagnosis of the unwinding process of a rolling mill. *Automation Technology in Practice (ATPi)*, pages 18–23, April 2006.

[8] Planchon, P., J. Lunze, M. Rode and M. Schneider. A cold rolling mill model for unwinding diagnosis. In *Proceedings of 4th MATHMOD Conference Series*, Vienna, February 2003.

[9] Puig, V., J. Quevedo, A. Stancu, J. Lunze, J. Neidig, P. Planchon, and P. Supavatanakul. Comparison of interval models and quantised systems in fault detection of the DAMADICS actuator benchmark problem. In *Proceedings of SAFEPROCESS*, Washington D.C., USA, 2003.

II. Further references

[10] Ackermann, J., editor. *Robust Control: Systems with Uncertain Physical Parameters.* Springer, London, U.K., 1993.

[11] Adrot, O. *Diagnostic à base de modèles incertains utilisant l'analyse par intervalles: l'approche bornante.* PhD thesis, Institut National Polytechnique de Lorraine, Centre de Recherche en Automatique de Nancy, 2000.

[12] Adrot, O., H. Janati-Idrissi and D. Maquin. Fault detection based on interval analysis. In *Proceedings of the 15th IFAC World Congress*, Barcelona, 2002.

[13] Alamo, T., J.M. Bravo and E.F. Camacho. Guaranteed state estimation by zonotopes. *Automatica*, 41:1035–1043, 2005.

[14] Basseville, M. On fault detectability and isolability. *European Journal of Control*, pages 625–637, 2001.

[15] Bay, J.S. *Fundamentals of Linear State Space Systems.* McGraw-Hill, 1999.

[16] Bentley, R.E. *The Theory and Practice of Thermoelectric Thermometry.* Springer, 1998.

[17] Bertsekas, D.P. and I.B. Rhodes. Recursive state estimation for a set-membership description of uncertainty. *IEEE Trans. on Automatic Control*, 16(2):117–128, 1971.

[18] Blanchini, F. Set invariance in control. *Automatica*, 35:1747–1767, 1999. Survey paper.

[19] Blanke, M., M. Kinnaert, J. Lunze and M. Staroswiecki. *Diagnosis and Fault-Tolerant Control.* Springer, Heidelberg, 2006.

[20] Boneh, A., S. Boneh and R.J. Caron. Constraint classification in mathematical programming. *Mathematical Programming*, 61:61–73, 1993.

[21] Boyd, S., L. El Ghaoui, E. Feron and V. Balakrishnan. *Linear Matrix Inequalities in System and Control Theory.* SIAM, 1994.

[22] Boyd, S. and L. Vandenberghe. *Convex Optimization.* Cambridge University Press, 2004.

[23] Brown, R.G. and P.Y. Hwang. *Introduction to Random Signals and Applied Kalman Filtering.* John Wiley & Sons, Inc., 1992.

[24] Bryant, G.F. *Automation of tandem mills.* The Iron and Steel Institute, 1973.

[25] Büeler, B., A. Enge and K. Fukuda. Exact volume computation for polytopes: A practical study. In *Computational Geometry*, 1996.

[26] Calafiore, G. Set simulations for quadratic systems. *IEEE Trans. on Automatic Control*, 48(5):800–805, May 2003.

[27] Caron, R.J., J.F. McDonald and C.M. Ponic. A degenerate extreme point strategy for the classification of linear constraints as redundant or necessary. *Journal of Optimization Theory and Applications*, 62:225–237, 1989.

[28] Chernousko, F.L. *State Estimation for Dynamic Systems.* CRC Press, 1994.

[29] Chernousko, F.L. Properties of optimal ellipsoids approximating reachable sets of uncertain systems. In 4th *IMACS Symposium on Mathematical Modelling*, pages 938–947, Vienna, February 2003.

[30] Chisci, L., A. Garulli and G. Zappa. Recursive state bounding by parallelotopes. *Automatica*, 32(7):1049–1055, 1996.

[31] Chow, E. and A. Willsky. Analytical redundancy and the design of robust failure detection systems. *IEEE Trans. on Automatic Control*, 29(7):603–614, 1984.

[32] Chung, D., C.G. Park and J.G. Lee. Robustness of controllability and observability of continuous linear time-varying systems with parameter perturbations. *IEEE Trans. on Automatic Control*, 44(10), October 1999.

[33] de Kleer, J.D. and B. Williams. Diagnosing multiple faults. *Artificial Intelligence*, 32:97–130, 1987.

[34] de Souza, C.E., U. Shaked and M. Fu. Robust H_∞ filtering for continuous time varying uncertain systems with deterministic input signals. *IEEE Trans. on Signal Processing*, 43(3):709–719, March 1995.

[35] Feldmann, F. *Modell Modifikation Band Coil-Rutscher (Funktionsbeschreibung).* ABB internal, ABB ASY/RDE, September 2004. Private communication.

[36] Förstner, D. *Qualitative Modellierung für die Prozessdiagnose und deren Anwendung auf Dieseleinspritzsysteme.* PhD thesis, Technische Universität Hamburg-Harburg, 2001.

[37] Frank, P.M. *Diagnosis in dynamical systems via state estimation – a survey*, volume 1. D. Reidel Publishing Company, 1987.

[38] Frank, P.M. Diagnoseverfahren in der Automatisierungstechnik. *Automatisierungstechnik*, 42:47–64, 1994.

[39] Frank, P.M. and X. Ding. Survey of robust residual generation and evaluation methods in observer-based fault detection systems. *Journal of Process Control*, 7(6):403–424, 1997.

[40] Geromel, J.C. and M.C. de Oliveira. H_2 and H_∞ robust filtering for convex bounded uncertain systems. *IEEE Trans. on Automatic Control*, 46(1):100–107, January 2001.

[41] Gertler, J.J. *Fault Detection and Diagnosis in Engineering Systems.* Marcel Dekker, 1998.

[42] Grünbaum, B. *Convex Polytopes.* Interscience Publishers, 1967.

[43] Guerra, P., J.M. Bravo, A. Ingimundarson, V. Puig and T. Alamo. Robust fault detection based on zonotope-based set-membership parameter consistency test. In *Proceedings of SAFEPROCESS*, Beijing, China, 2006.

[44] Guerra, P., V. Puig, A. Ingimundarson and M. Witczak. Robust fault detection with unknown input set-membership state estimator and interval models using zonotopes. In *Proceedings of SAFEPROCESS*, Beijing, China, 2006.

[45] Hamelin, F. and T. Boukhobza. Geometric-based approach to fault detection for multilinear affine systems. In *Proceedings of the 16th IFAC World Congress*, Prag, 2005.

[46] Hamelin, F., H. Noura and D. Sauter. Fault detection method of uncertain system using interval model. In *European Control Conference*, pages 826–831, Porto, 2001.

[47] Hamscher, W., L. Console and J. de Kleer. *Readings in Model-Based Diagnosis.* Morgan Kaufmann, 1992.

[48] Harris, T.J., C.T. Seppala and L.D. Desborough. A review of performance monitoring and assessment techniques for univariate and multivariate control systems. *Journal of Process Control*, 9:1–17, 1999.

[49] Heeks, J., E.P. Hofer, B. Tibken and K. Thorwart. An interval arithmetic approach for discrete time nonsmooth uncertain systems with application to an antilocking system. In *American Control Conference*, pages 1849–1854, Anchorage, 2002.

[50] Holzweissig, F. and H. Dresig. *Lehrbuch der Maschinendynamik*. Springer-Verlag, 1982.

[51] Horch, A. *Condition Monitoring of Control Loops*. PhD thesis, Royal Institute of Technology, Stockholm, Sweden, 2000.

[52] iba AG. `http://www.iba-ag.com`.

[53] Ingimundarson, A., J.M. Bravo, V. Puig and T. Alamo. Robust fault diagnosis using parallelotope-based set-membership consistency tests. In *European Control Conference and IEEE Conference and Decision and Control*, pages 993–998, Seville, 2005.

[54] Isermann, R. Beispiele für die Fehlerdiagnose mittels Parameterschätzung. *Automatisierungstechnik*, 37(9):336–343, 1989.

[55] Isermann, R. Modellgestützte überwachung und Fehlerdiagnose Technischer Systeme, Teil 1. *Automatisierungstechnische Praxis*, 5:9–20, 1996.

[56] Isermann, R. Modellgestützte überwachung und Fehlerdiagnose Technischer Systeme, Teil 2. *Automatisierungstechnische Praxis*, 6:48–57, 1996.

[57] Jaulin, L., M. Kieffer, O. Didrit and E. Walter. *Applied Interval Analysis*. Springer-Verlag, London, 2001.

[58] Jia, Y. Robust control with decoupling performance for steering and traction of 4WS vehicles under velocity-varying motion. *IEEE Trans. on Control Systems Technology*, 8(3):554–569, May 2000.

[59] Jolliffe, I.T. *Principal Component Analysis*. Springer-Verlag, New-York, 1986.

[60] Kailath, T. *Linear Systems*. Information and System Sciences Series. Prentice Hall International, Inc., London, 1980.

[61] Kalman, R.E. and R.S. Bucy. New results in linear filtering and prediction theory. *Trans. of the ASME, Series D, Journal of Basic Engineering.*, 83:95–100, 1961.

[62] Karwan, M.H., V. Lotfi, J. Telgen and S. Zionts, editors. *Redundancy in Mathematical Programming.* Springer, Berlin, 1983.

[63] Keerthi, S.S. and E.G. Gilbert. Computation of minimum-time feedback control laws for discrete-time systems with state-control constraints. *IEEE Trans. on Automatic Control*, 32(5):432–435, May 1987.

[64] Kerrigan, E.C. *Robust Constraint Satisfaction: Invariant Sets and Predictive Control.* PhD thesis, University of Cambridge, November 2000. Available online.

[65] Koeppen, B. *Analyse und Abschätzung der Beobachtungsgenauigkeit von Intervallbeobachtern.* Diplomarbeit, Technische Universität Hamburg-Harburg, Januar 2001.

[66] Kuipers, B.J. *Qualitative reasoning. Modeling and simulation with incomplete knowledge.* The MIT Press, 1994.

[67] Kurzhanski, A. and I. Vályi. *Ellipsoidal Calculus for Estimation and Control.* Birkäuser, 1996.

[68] Kvasnica, M., P. Grieder and M. Baotić. *Multi-Parametric Toolbox (MPT).* `http://control.ee.ethz.ch/~mpt/`, 2004.

[69] Lalami, A. and C. Combastel. Generation of set membership tests for fault diagnosis and evaluation of their worst case sensitivity. In *Proceedings of SAFEPROCESS*, Beijing, China, 2006.

[70] Lamperti, G. and M. Zanella. *Diagnosis of Active Systems.* Kluwer Academic Publishers, Dordrecht, Netherlands, 2003.

[71] Lin, F. Diagnosability of discrete-event systems and its applications. *Journal of Discrete Event Dynamic Systems*, 4(2):197–212, 1994.

[72] Lin, H. and P.J. Antsaklis. *A Necessary and Sufficient Condition for Robust Asymptotic Stabilizability of Continuous-Time Uncertain Switched Linear Systems.* Technical Report 2004-002, ISIS, 2004. Available online.

[73] SysBrain Ltd. *The Geometrical Bounding Toolbox (GBT) for Matlab.* `http://sysbrain.com/`.

[74] Luenberger, D.G. An introduction to observers. *IEEE Trans. on Automatic Control*, 16:596–603, 1971.

[75] Lunze, J. *Künstliche Intelligenz für Ingenieure, Band 2.* Oldenbourg, München, 1995.

[76] Lunze, J. *Automatisierungstechnik.* Oldenburg, 2003.

[77] Lunze, J. *Ereignisdiskrete Systeme.* Oldenbourg, 2006.

[78] Lunze, J. *Regelungstechnik, Band 1.* Springer, 2006.

[79] Lunze, J. *Regelungstechnik, Band 2.* Springer, 2006.

[80] Lunze, J., J. Chen, P.M. Frank, M. Kinnaert and R.J. Patton. *Control of Complex Systems*, chapter "Fault Detection and Isolation", pages 191–207. Springer, 2001.

[81] Maksarov, D.G. and J.P. Norton. State bounding with ellipsoidal set description of the uncertainty. *International Journal of Control*, 65(5):847–866, 1996.

[82] Mangoubi, R.S. *Robust estimation and failure detection.* Springer-Verlag, 1998.

[83] The Mathworks. *Getting Started with MATLAB.* The MathWorks, Inc., 2005.

[84] The Mathworks. *Getting Started with Simulink.* The MathWorks, Inc., 2005.

[85] Milanese, M. and A. Vicino. Optimal estimation theory for dynamic systems with set membership uncertainty: An overview. *Automatica*, 27(6):997–1009, 1991.

[86] Millerioux, G., L. Rosier, G. Bloch and J. Daafouz. Bounded state reconstruction error for LPV systems with estimated parameters. *IEEE Trans. on Automatic Control*, 49(8):1385–1389, August 2004.

[87] Neidig, J. and J. Lunze. Unidirectional coordinated diagnosis of automata networks. In *Proceedings of the 17th International Symposium MTNS*, pages 2203–2208, Kyoto, 2006.

[88] O'Reilly, J. *Observers for linear systems.* Academic Press, 1983.

[89] Patton, R.J., P.M. Frank and R.N. Clark. *Issues of Fault Diagnosis for Dynamic Systems.* Springer, 2000.

[90] Ploix, S. and O. Adrot. Parity relations for linear dynamic systems with multiplicative uncertainties. In *Proceedings of SAFEPROCESS*, Beijing, China, 2006.

[91] Polyak, B.T., S.A. Nazin, C. Durieu and E. Walter. Ellipsoidal parameter or state estimation under model uncertainty. *Automatica*, 40:1171–1179, 2004.

[92] Puig, V., A. Ingimundarson and S. Tornil. Robust fault detection using inverse images of interval functions. In *Proceedings of SAFEPROCESS*, Beijing, China, 2006.

[93] Puig, V., J. Quevedo, T. Escobet and S. de las Heras. Passive robust fault detection approaches using interval models. In *Proceedings of the 15th IFAC World Congress*, Barcelona, July 2002.

[94] Puig, V., J. Quevedo, T. Escobet and A. Stancu. Robust fault detection using linear interval observers. In *Proceedings of SAFEPROCESS*, Washington D.C., USA, 2003.

[95] Puig, V., A. Stancu and J. Quevedo. Passive robust fault detection using a forward-backward test. In *Proceedings of SAFEPROCESS*, Beijing, China, 2006.

[96] Pulido, B., J.J. Rodriguez Diez, C. Alonso Gonzalez, O.J. Prieto and E.R. Gelso. Diagnosis of continuous dynamic systems: Integrating consistency based diagnosis with machine-learning techniques. In *Proceedings of the 16th IFAC World Congress*, Prag, 2005.

[97] Sampath, M., R. Sengupta, S. Lafortune, K. Sinnamohideen and D. Teneketzis. Diagnosability of discrete-event systems. *IEEE Trans. on Automatic Control*, 40:1555–1575, September 1995.

[98] Savkin, A.V. and I.R. Petersen. Robust state estimation and model validation for discrete-time uncertain systems with a deterministic description of noise and uncertainty. *Automatica*, 34(2):271–274, 1998.

[99] Savkin, A.V. and I.R. Petersen. Weak robust controllability and observability of uncertain linear systems. *IEEE Trans. on Automatic Control*, 44(5), May 1999.

[100] Sayed, A.H. A framework for state-space estimation with uncertain models. *IEEE Trans. on Automatic Control*, 46(7):998–1013, July 2001.

[101] Schild, A. *Anwendung einer Diagnosemethode mit qualitativem Modell an einem Druckluftkraftwerk.* Diplomarbeit, Ruhr-Universität Bochum, Lehrstuhl für Automatisierungstechnik und Prozessinformatik, 2004.

[102] Schröder, J. *Modelling, State Observation and Diagnosis of Quantised Systems.* Springer, 2003.

[103] Schweppe, F.C. Recursive state estimation: unknown but bounded errors and system inputs. *IEEE Trans. on Automatic Control*, AC-13(1):22–28, February 1968.

[104] Schweppe, F.C. *Uncertain Dynamic Systems.* Prentice Hall, 1973.

[105] Shamma, J.S. and K.-Y. Tu. Approximate set-valued observers for nonlinear systems. *IEEE Trans. on Automatic Control*, 42(5):648–658, May 1997.

[106] Staroswiecki, M. and G. Comtet-Varga. Fault detectability and isolability in algebraic dynamic systems. In *European Control Conference*, Karlsruhe, 1999.

[107] Supavatanakul, P. *Modelling and diagnosis of timed discrete-event systems.* PhD thesis, Ruhr-Universität Bochum, 2004.

[108] Telgen, J. Minimal representation of convex polyhedral sets. *Journal of Optimization Theory and Applications*, 38(1):1–24, September 1982.

[109] Tibken, B. and E.P. Hofer. A new simulation tool for uncertain discrete time systems. In *European Control Conference*, pages 814–817, Groningen, The Netherlands, 1993.

[110] Tiller, M.M. *Introduction to Physical Modeling with Modelica.* Kluwer Academic, 2001.

[111] Travé-Massuyès, L., M.-O. Cordier and X. Pucel. Comparing diagnosability in CS and DES. In *Proceedings of SAFEPROCESS*, Beijing, China, 2006.

[112] Carnegie Mellon University. *Checkmate's Hybrid System Verification Toolbox for Matlab.* `http://www.ece.cmu.edu/~webk/checkmate/`.

[113] Ushio, T., I. Onishi and K. Okuda. Fault detection based on petri net models with faulty behaviors. In *IEEE International Conference on Systems, Man and Cybernetics*, volume 1, pages 113–118, San Diego, 1998.

[114] Venkatasubramanian, V., R. Rengaswamy, K. Yin and S.N. Kavuri. A review of process fault detection and diagnosis. Part I: Quantitative model-based methods. *Computers and Chemical Engineering*, 27:293–311, 2003.

[115] Venkatasubramanian, V., R. Rengaswamy and S.N. Kavuri. A review of process fault detection and diagnosis. Part II: Qualitative models and search strategies. *Computers and Chemical Engineering*, 27:313–326, 2003.

[116] Venkatasubramanian, V., R. Rengaswamy, S.N. Kavuri and K. Yin. A review of process fault detection and diagnosis. Part III: Process history based methods. *Computers and Chemical Engineering*, 27:327–346, 2003.

[117] Wang, K. and A.N. Michel. Necessary and sufficient conditions for the controllability and observability of a class of linear time-invariant systems with interval plants. *IEEE Trans. on Automatic Control*, 39(7):1443–1447, July 1994.

[118] Wang, Z. and H. Unbehauen. Robust H_2/H_∞-state estimation for discrete-time systems with error variance constraints. *IEEE Trans. on Automatic Control*, 42(10):1431–1435, October 1997.

[119] Willems, J.C. Paradigms and puzzles in the theory of dynamical systems. *IEEE Trans. on Automatic Control*, 36:259–294, March 1991.

[120] Witsenhausen, H.S. Sets of possible states of linear systems given perturbed observations. *IEEE Trans. on Automatic Control*, 13:556–558, October 1968.

[121] Xie, L., C.E. de Souza and M. Fu. H_∞ estimation for discrete-time linear uncertain systems. *International Journal of Robust and Nonlinear Control*, 1, 1991.

[122] Ziegler, G.M. *Lectures on Polytopes.* Springer-Verlag, 1995.

Resume

Name:	Philippe Planchon
Birthday:	March 22nd, 1978
Birthplace:	London, United Kingdom

School education

1983 – 1992	Ecole and Collège Guynemer, Compiègne, France
1992 – 1995	Lycée Saint-Dominique, Neuilly-sur-Seine, France

College education

1995 – 1997	Preparatory College – Classes Préparatoires "Math-Sup" at Lycée Pasteur, Paris, France "Math-Spé" at Lycée Buffon, Paris, France
1997 – 2000	Université de Technologie de Compiègne, France Degree of Mechanical Engineering
1999 – 2000	University of Illnois at Urbana-Champaign, USA Electrical and Computer Engineering – Exchange program

Work experience

1996 – 1997	Robert Bosch GmbH Internship in factory of Stuttgart-Feuerbach, Germany
2000 – 2000	Asea Brown Boveri (ABB) Internship at ABB Corporate Research Center, Heidelberg, Germany
2001 – 2001	Renault S.A. Postgraduate internship in Technocentre, Guyancourt, France
2001 – 2006	Ruhr-Universität Bochum Research Assistant Institute of Automation and Computer Control
2007 –	BMW Group Service processes and vehicle diagnosis
